6116

DRAINING CHICAGO

The Early City and the North Area

D1599180

Richard Lanyon

Draining Chicago: The Early City and the North Area

Richard Lanyon

Published April 2016 by Richard Lanyon

Distributed by:

Lake Claremont Press: A Chicago Joint, an imprint of Everything Goes Media, LLC
www.lakeclaremont.com
www.everythinggoesmedia.com

Copyright © 2016 by Richard Lanyon

All rights reserved. No part of this book may be reproduced or transmitted in any form or by any means, electronic or mechanical, including photocopying, recording, or by any information storage or retrieval system without written permission from the publisher, except for the inclusion of brief quotations in a review.

Publisher's Cataloging-In-Publication Data

(Prepared by The Donohue Group, Inc.)

Names: Lanyon, Richard.

Title: Draining Chicago : the early city and the North Area / Richard Lanyon.

Description: [Chicago, Illinois] : Richard Lanyon, 2016. | [Chicago, Illinois] : Lake Claremont Press, a Chicago Joint, an imprint of Everything Goes Media, LLC | Includes bibliographical references and index.

Identifiers: LCCN 2015961037 | ISBN 978-1-893121-73-7 |

ISBN 978-1-893121-70-6 (ebook)

Subjects: LCSH: Public works--Illinois--Chicago--History. | Drainage--Illinois--Chicago--History. | Sewerage--Illinois--Chicago--History. | Water diversion--Illinois--Chicago River--History. | Chicago (Ill.)--History. | North Chicago (Ill.)--History.

Classification: LCC F548.3 .L36 2016 | LCC F548.3 (ebook) | DDC 977.3/11--dc23

20 19 18 17 16 10 9 8 7 6 5 4 3 2 1

Author inscribed copies available at www.lakeclaremont.com.

R0445499719

To my sister Ellen Lanyon, internationally known visual artist, whose untimely death in October 2013 occurred while I was preparing this manuscript. I'm ever grateful for her inspiration and guidance.

Contents

CHAPTER 3: NORTH SHORE CHANNEL

CHAPTER 4: NORTH AREA SEWERSHEDS AND WATERSHEDS

APPENDICES

List of Tables

List of Figures

List of Photographs

CHAPTER 2: NEW DIGS FOR THE NORTH BRANCH

CHAPTER 3: NORTH SHORE CHANNEL

CHAPTER 4: NORTH AREA SEWERSHEDS AND WATERSHEDS

APPENDICES

Preface

The daily flushes from five million people, lots of other building drainage, and copious amounts of water from the sky—where does it all go? Let the story begin.

My first book, *Building the Canal to Save Chicago*, was a heroic tale of a world class public works project. Flooding and public health problems were devastating, and a plan for the City of Chicago (City) had to be developed and implemented. A deep waterway connection between the Great Lakes and Mississippi River was also being promoted and the promoters saw opportunity. A new entity was created, the Sanitary District of Chicago (District), and the promoters of the deep waterway and City leaders combined forces to execute the plan against many odds. The elected trustees and engineers of the District, as well as the contractors and their many employees, were the players in this heroic tale, using their sense of mission and their ingenuity, know-how, and plain hard work to get the immediate job—the Canal—done.

But a greater job remained. Simply reversing the flow of the Chicago River and South Branch to remove the odoriferous nuisance and keep the river from polluting Lake Michigan, the city's water supply, wasn't enough. There were sewers discharging directly to Lake Michigan and another river, the Calumet, threatening the water supply. The North Branch was severely polluted and growing worse as the city's North Side developed. The Calumet River was becoming more polluted from the growing industrial and residential area on the city's far South Side. Growing suburbs along the lakefront to the north were also discharging sewage to the lake. The South Branch remained narrow and congested with insufficient capacity to convey the statutorily required dilution water from the lake to the new Sanitary & Ship Canal. Growing communities without direct access to the lake or a river needed an outlet for their volumes of sewage and stormwater.

To preserve and protect public health, good drainage is fundamental. For the Chicago area, the need for drainage was crucial in this once natural, flat, and poorly drained landscape populated with marshes, prairies, wetlands, and woodlands. The land surface was young in geologic time, without a well-developed network of rivers and streams. Chicago was growing rapidly, population increasing through the influx of industry and people, land area growing through annexations of adjacent areas and communities. The District was faced with its mission of protecting the water supply while dealing with a rapid expansion of the city and surrounding suburbs. In sum, at the start of the twentieth century the District faced many challenges.

My first book left the District at the turn of the century and it was my initial intention to continue the history of the District with a volume covering each decade. With my rudimentary knowledge of the District's history, I thought this logical. However, after a year of historical research dwelling on the first decade, I found myself with several incomplete stories to tell. I also found that my initial rudimentary knowledge of the District's history had been misleading. I learned, to my surprise, that the District was in the midst of building 14 bridges over the Chicago River and South Branch, the North Shore Channel was not completed by 1910 as believed, intense focus on the generation and distribution of hydroelectric energy was interfering with the District's mission to protect public health, and construction of the Calumet-Sag Channel was held up by controversy and litigation.

There would be no book on the first decade. Instead, as I continued with more historical research to better my understanding, it became clear to me that the District's prime responsibility was providing an outlet for draining the Chicago area. Before the District, the City raised the streets and installed sewers. Later, the City and District cooperated in the elimination of sewage discharges to Lake Michigan. To tell the story of drainage I realized that I had to go back to the 1850s when the City began systematic sewer construction.

This book tells of that early work, then focuses on three areas: the northeast area of Cook County, the North Branch watershed, and the O'Brien Water Reclamation Plant sewershed. *Sewershed* is the term used to describe the replacement of natural drainage by man-made drainage systems of sewers, water reclamation plants, deep tunnels, etc. Subsequent books will deal with the continuing build-out of other parts of the canal system such as the Calumet and Stickney Water Reclamation Plant sewersheds and will describe lake diversion litigation and the transition from sewage dilution to sewage treatment.

Reversing the flow of a river can only be accomplished when the topography makes it possible. I'm often asked how it was accomplished. Reversing the direction of flow of a river is not the same as transferring water from one watershed to another. The transfer of water, a common practice, is often done for irrigation or hydroelectric power generation. The flow of the river involved is not reversed in its course; rather, it is diverted and gravity in canal, pipeline, or tunnel, or pumping makes the water flow.

The Chicago area's location on a subcontinental divide is unique. The Midwest is a horseshoe-shaped watershed surrounding the Mississippi River valley. Pretend you are a giant standing at the open end of the horseshoe in New Orleans. On a clear day the Rocky Mountains rise to your left in the west and the Appalachians rise to your right in the east. Straight ahead to the north, but lower than either aforementioned mountain range, is the upland plateau of the northern Great Plains. A little to the right are the Great Lakes, huge bathtubs sunk deep into the land surface. The nearest, Lake Michigan, is just beyond the reach of the Illinois River, which branches off the mighty Mississippi and courses north, then northeast, across the plains of Illinois. The Chicago area sits in a shallow depression on the southwest rim of Lake Michigan. Much of that depression is the Chicago Lake Plain, a flat expanse slightly above the lake surface. The Chicago River courses through the flat plain and empties into the lake.

That topographic setting made reversing the flow of the river possible: simply breach the southwest rim of the depression and the lake will flow into the Illinois River valley. Modern geologists studied and explained the glacial history of the Chicago region and Lake Michigan in the twentieth century, but early explorers such as Louis Jolliet and Père Marquette knew what was possible by canoeing the rivers and walking the land. Later, engineers Ellis Chesbrough, Lyman Cooley, and Rudolph Hering benefitted from government land surveys and topographic maps to understand the lay of the land. Today, the Illinois Waterway is the continuous navigable link between the Mississippi River and Lake Michigan, subsuming the Chicago area canals, lower Des Plaines River, and Illinois River.

This book, enriched with photographs of the era, is a first step in revealing the history of drainage in the Chicago area. The reader will be taken through the nineteenth-century efforts of the City in planning and implementing a sewer system; follow the work of the City and District in eliminating sewer discharges to the lake; and learn about the District's improvement of the North Branch, the construction of the North Shore Channel, and the building of sewers and sewage treatment plants. The discussion of stormwater management, including the Deep Tunnel, will bring the history of drainage up-to-date for the *north area*. The north area, which refers to the sewershed imposed on the landscape by man-made drainage infrastructure, is defined generally by the Cook-Lake County border on the north, the lakefront on the east, Fullerton Avenue on the south, and Harlem Avenue and the Des Plaines River on the west. The west border, somewhat ambiguous, will be better defined later.

This book makes frequent use of three acronyms: CCD, Chicago City Datum, the reference level for elevations used in the Chicago area set in 1847 as low water of Lake Michigan; cfs, cubic feet per second, a measure of the rate of water flow; and gpm, gallons per minute, another measure of water flow. CCD is second nature to many in the Chicago area who deal with infrastructure. For the other two, cfs is commonly used for streamflow records by the U.S. Geological Survey; and gpm, with its familiar unit of gallons, is universally used in the field of water resources. The railroad and street names are those in use at the time, with current names occasionally indicated. The chapters in this book describe drainage in the early city via canals and sewers. Along the way are references to some related topics; those related topics appear in appendices.

References are listed at the end of each chapter, but the narrative avoids superscripts and footnotes. Much of what is written is based on what I've absorbed over 50 years working in the water business. My first remembered experience was at the age of ten when my family moved into a home on Giddings Street in the city's Ravenswood Manor neighborhood. The house was two doors from the North Branch; my bedroom window looked out on the channel, the Lawrence Avenue Bridge, and North Branch Pumping Station. The attraction of water, trees, and channel made the North Branch an adventurous playground. My District-related experience, which began as an undergraduate at the University of Illinois in Urbana, was working in the hydraulic engineering laboratory on a physical model study of a discharge chamber at the Lockport powerhouse.

Except for George P. Brown in 1892, no historian has documented the District's history. Brown's history, published in 1894, is rich with detail of the events leading up to the start of the construction of the Sanitary & Ship Canal. It is my hope to help readers understand the history of the District and its works. My telling is entirely voluntary because I stand in awe of the District and its accomplishments. Without the work of the District, the Chicago metropolis could not exist.

Public agencies, the District included, present their best face via news releases and other outlets. Judged from the outside, some information releases are more successful than others. Perhaps it is because the news media usually focuses on the failures, scandals, and tragedies that the agencies themselves focus on the successes: awards, groundbreakings, innovations, milestones, and ribbon cuttings. The built environment and infrastructure are vital for urban livability, and their stories need to be told.

As far as I have been able to determine, the stated history of the works of the District herein is accurate and factual. All opinions expressed are mine and should not be construed as District policy.

Prologue

Chicago Lake Plain

The rim of the Great Lakes watershed is close to a lake shore only in Chicago. And the vertical lift to surmount the divide was only several feet. Truly, a gift of the glaciers. When European settlers arrived in the early 1800s the average distance from lake shore to divide was about eight miles as the crow flies, perhaps ten miles by water. The marsh area surrounding Mud Lake shrunk to a length of three miles in a dry year and expanded to six miles in a wet year. On the other side of the divide, the Des Plaines River was about ten miles from the shoreline; it was several feet above the lake in dry times, more in wet times. If the succession of glacial retreat had been different, the Des Plaines River might still have been flowing into Lake Michigan when Europeans arrived. As Illinois became a state (1818) and the settlers turned Chicago into a village (1833) and then a city (1837), this topographical detail was turned into opportunity. The Illinois & Michigan Canal opened in 1848, breaching the divide to create an avenue of water commerce. The city, already a busy port on Lake Michigan, became a canal town. (See Figure 1.)

However, when the Des Plaines River went into flood mode and forced some of its excess across the divide to cause havoc and damage in the city, the topographical opportunity became a threat. A severe flood in 1849 swept boats and bridges out into Lake Michigan and caused extensive damage to riparian property along the West Fork, South Branch, and Chicago River. In 1876, near what is now 4900 South

Harlem Avenue, the City constructed Ogden Dam across Portage Creek to hold back the Des Plaines River. The dam successfully retarded smaller floods, but another mega-flood in 1885 wiped out the dam and swept the river clean, depositing much debris and filth in the lake. While this did not, as some believe, cause a major epidemic of waterborne disease, it certainly became clear that the hydrologic and hydraulic vagary, combined with the perennial wetness of the marsh and prairie landscape, had to be overcome for the city to grow. Railroads came along shortly after the Illinois & Michigan Canal opened, adding to the growth of the city and placing more pressure on City leaders to drain the land.

Water Supply and Sewage

Beginning in 1840, the Chicago Hydraulic Company supplied water obtained from the lake by pumping through wooden water mains; the City took over the operation in 1851. Sewer construction began in the late-1850s with outfalls to the river and lake. Adding the discharge of sewage worsened the public health situation, already aggravated by poor drainage and backyard privies; sewage flowing into the lake was polluting the water supply.

Extending the water supply intake tunnel two miles from shore in 1867 helped reduce the threat to the water supply, but didn't eliminate the threat. Pumping water from the South Branch into the Illinois & Michigan Canal was attempted to reduce the flow to the lake. Executing the deep cut plan for the Illinois & Michigan Canal proved insufficient. The growing volume of sewage and added waste from the Stock Yards and many other industries was simply overwhelming. The Chicago Fire in 1871 diverted attention from the river while the city recovered, but by 1880 the citizens were again agitated by the public health issue. The public water supply was plagued with inadequate pressure at times, even with more water tunnels and pumping stations. Without any treatment, water from spigots contained whatever could make its way through the water mains.

In 1858, Ellis S. Chesbrough, from Boston, the first chief engineer of the newly formed Chicago Board of Sewerage Commissioners, suggested that a canal larger than the Illinois & Michigan (I&M) Canal would easily reverse the flow of sewage into the lake. Despite stopgap efforts to increase the capacity of the I&M Canal, Chesbrough's suggestion languished for 30 years.

Despite citizen protest and request, the City was not moved to be daring until the devastating storm of 1885 reminded all that something daring was needed. The Chicago Commission on Drainage and Water Supply was created and hired Rudolph Hering of Philadelphia to recommend a solution. Hering fleshed out the concept of a larger canal with specific dimensions, a dilution ratio, and a cost estimate. Hering's design took hold—in 1887 City leaders determined that it was essential to change the flow of the river away from the lake by a canal cut deep into the limestone bedrock. A distance of only 28 miles would connect the South Branch to the Des Plaines River at Lockport, a place where the bedrock surface was declining toward Joliet at the rate of about eight feet per mile.

The Sanitary District of Chicago was created in 1889, a new form of a special purpose district, to implement the Hering plan to reverse the flow of the Chicago River and thus keep sewage out of the lake and protect the water supply. Hering also recommended measures, which the City readily adopted, to improve the City water distribution system. Canal building by the District proceeded and by 1900 the flow of the Chicago River and South Branch was in the opposite direction.

Thinking Ahead

As the turn of the century approached, the District realized that it was about to achieve the objective of reversing the flow of the Chicago River. However, it also realized that to achieve the purpose of the Act of 1889—to protect the water supply—much more needed to be accomplished. The intent of the act had been framed in the late-1880s, and much had changed in a decade. The city continued to grow rapidly in area and population, exceeding expectations. The Calumet area, with its industrial and residential growth, was annexed into the city. The Calumet River was becoming as great a threat to the water supply as the Chicago River had been. North suburban communities were growing and their sewers were polluting the lake. Neighborhoods in the city and suburban communities, some far from the lake or a river, needed outlets for their drainage and sewage. The federal government was using the commerce clause of the U.S. Constitution to increase its authority over navigable waters. Pressure continued for a deep waterway connection between the Great Lakes and Gulf of Mexico, but the federal government was not ready to support such a water resource development.

The District, in cooperation with the City and suburbs, had to be creative in using its statutory authority in pursuit of its mission to protect the water supply. The reversal of flow in the Chicago River and South Branch, a significant achievement, would begin a new era for the District. But there were many more problems to address, priorities were difficult to establish, and tax resources were limited. The trustees had ideas about what was needed and the engineers were developing plans, but it was premature to go public until the Sanitary & Ship Canal was completed, the flow of the river was reversed, and the impatient citizenry were satisfied.

And satisfied they were, almost immediately. Both the *Chicago Tribune* and *New York Times* reported on January 4, 1900, that a miracle had occurred; the Chicago River was running clear—in the other direction—and fish were observed that hadn't been seen since the Chicago Fire in 1871. The miracle happened just from filling the 28-mile long excavation. A constant flow of about 500 cfs in the middle of winter would easily make that impression. The flow abated on January 14 because the gates at Lockport were closed and the excavation was full. What if Governor Tanner didn't approve the opening of the gates at Lockport? Thankfully he did, and on January 17 the gate was opened and the flow was nearly ten times what it was the prior week, meeting the dilution requirement in the Act of 1889. In a few years the decrease in mortality statistics would show the public health benefit, but in the meantime it was necessary to fulfill the District's mission.

The April 1900 Plan and the Special Commission of Experts

The early 1900s was a time without proven technologies for large-scale sewage treatment. It was known that flowing rivers purified themselves and the District used this knowledge, along with the concept of dilution, to flush urban sewage away from the source of the water supply. Dilution wasn't unique to the District; it was practiced in other big cities. However, in most other big cities the source was upstream and the discharge was downstream; in the Chicago area, until the Sanitary & Ship Canal was opened, the source and discharge were one and the same, Lake Michigan.

Drawing together legal opinions, assessments of financial capability, numerous plans for improvement prepared by the engineering department, and, most importantly, the recommendations of Chief Engineer Isham Randolph, the April 1900 plan laid out a path forward. The plan was a framework of options to address sewage from the northern lakefront suburbs, an outline of cooperation with the City in eliminating the sewers discharging directly to the lake, proposals for hydraulic capacity and navigation improvement for the South Branch, criteria for development of hydroelectric generation at Lockport, and alternatives for a channel to reverse the flow of the Calumet River. Randolph included estimated costs, which were staggering. The trustees, hesitant to move forward because of the financial obligations involved and the lack of confidence that all the work outlined in the plan was necessary, called for a second opinion.

In January 1901, in a major departure for the trustees, they looked for advice outside the District and designated the Special Commission of Experts. The trustees tasked the commission to investigate and recommend projects for improving the South Branch—bridge replacement, channel widening, new dock walls, and dredging. The commission sought public input and, with the assistance of the engineering department, assembled a large amount of information, plans, and studies. Three of the commission members were engineers and the others were attorneys; all were familiar with the work of the District, but were not previously affiliated with District work or other governmental agencies working with the District.

Mr. Randolph was influential with the experts and the experts respected and trusted him. The April 1900 plan was sufficient for the chief engineer, and the commission confirmed the plan. Although the commission went beyond its charge, its report didn't materially alter the course of the District. The trustees were satisfied because the public appeared to support the plan. The next step was to obtain necessary legislation to annex areas to the District, to authorize what needed to be accomplished, and to obtain additional bonding authority.

To understand the District's sense of mission at the close of the nineteenth century as to what had to be accomplished, the next chapter steps back in time to 1855 when the City addressed the need for drainage and began the systematic construction of sewers.

References

Brown, George P. *Drainage Channel and Waterway: A History of the Effort to Secure an Effective and Harmless Method for the Disposal of the Sewage of the City of Chicago, and to Create a Navigable Channel Between Lake Michigan and the Mississippi River*. Chicago: R.R. Donnelley & Sons Company, 1894.

Chicago Department of Public Works. *A Century of Progress in Water Works, 1833–1933*. Chicago: City of Chicago, 1933.

Chicago Department of Public Works. *The Chicago Sewer System: 100 Years of Protecting Chicago's Health*. Chicago: City of Chicago, 1956.

Chicago Tribune Archives. archives.chicagotribune.com.

Knight, Robert, and Lucius H. Zeuch. *The Location of the Chicago Portage Route of the Seventeenth Century*. Chicago: Chicago Historical Society, 1928.

Lanyon, Richard. *Building the Canal to Save Chicago*. Bloomington, IN: Xlibris Corporation, 2012.

Sanitary District of Chicago. Proceedings of the Board of Trustees, 1900 and 1901. Chicago.

New York Times Archives. nytimes.com/ref/membercemtermutarchive.html.

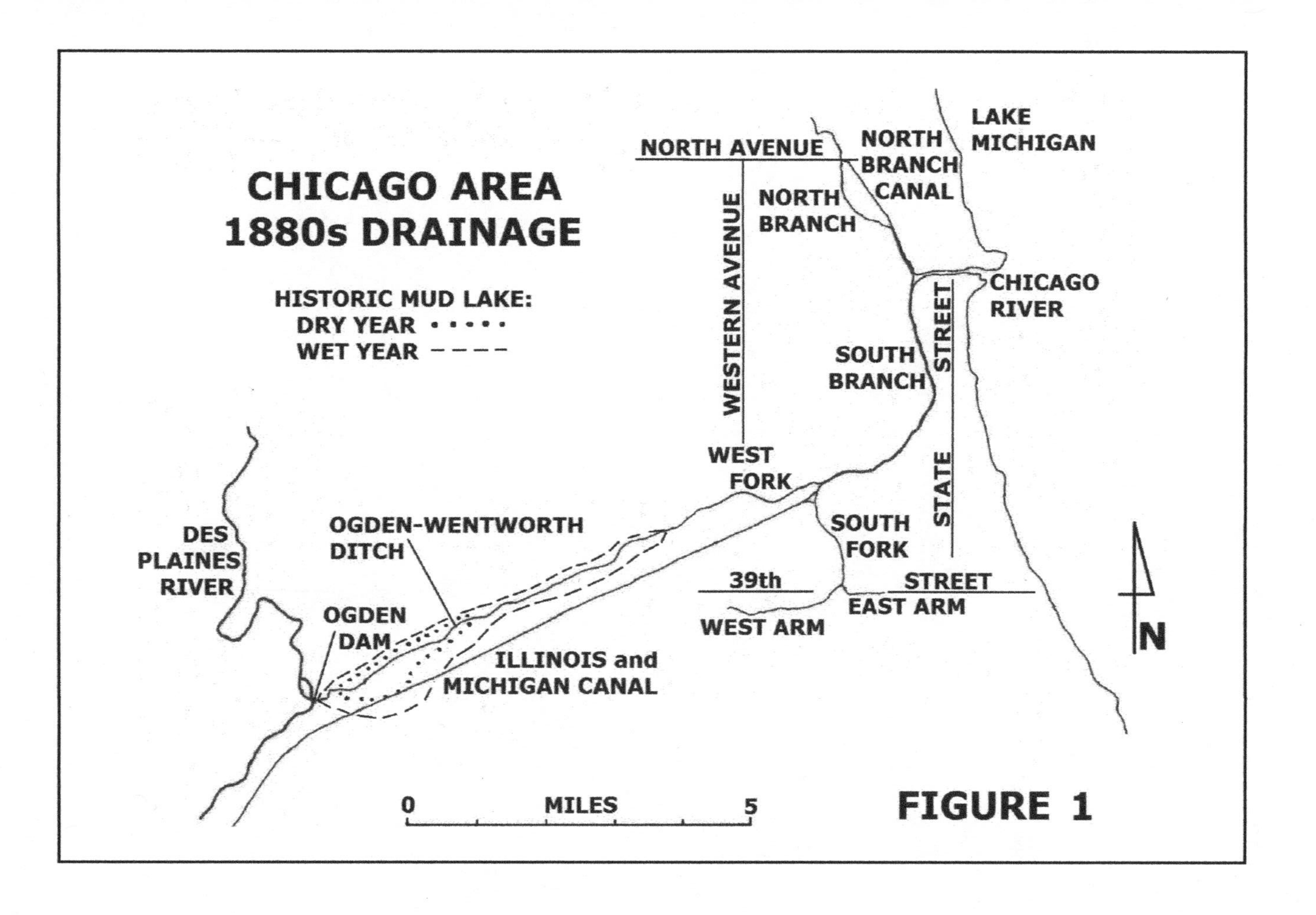
CHICAGO AREA
1880s DRAINAGE
HISTORIC MUD LAKE:
DRY YEAR • • • • •
WET YEAR — — — —
NORTH AVENUE
NORTH BRANCH CANAL
LAKE MICHIGAN
NORTH BRANCH
WESTERN AVENUE
CHICAGO RIVER
SOUTH BRANCH
STATE STREET
WEST FORK
SOUTH FORK
39th STREET
EAST ARM
WEST ARM
DES PLAINES RIVER
OGDEN-WENTWORTH DITCH
OGDEN DAM
ILLINOIS and MICHIGAN CANAL
N
0
MILES
5

FIGURE 1

Chapter 1

Sewers for Chicago: 1855–1900

Introduction

Amazingly, people in Chicago survived without sewers for four decades. You wouldn't expect the early explorers or hunters to care; after all, they didn't put down roots. Even the first settlers with cabins here and there on the landscape along the Chicago River could get by without sewers. By the time Chicago became a village in 1833 with about 200 residents, it was still a frontier settlement in the midst of a marsh, prairie, and woodland landscape. Three years after Chicago was chartered as a city, the 1840 census was about 4,000 and still no sewers. Streets were not paved and roadside ditches sufficed to carry away excess water to the Chicago River or its two branches. Sewage was managed in privies, pit or otherwise. It wasn't pollution of Lake Michigan that first drew attention to the need for sewers in the city; rather, the problem was poor drainage of the land: stagnant water accumulated in low areas breeding mosquitos and emitting foul odors.

Cholera was rampant in the mid-1850s and even though belief in miasma was popular, the germ theory being decades into the future, intuition told many that sewage and human wellbeing were incompatible. Sewage was putrescible and gave off disgusting odors, feeding those who believed in miasma. The city had one of the highest mortality rates in the world in the late 1800s, with cholera earlier and typhoid later as leading causes of death.

Sewer construction became popular in some major cities in the 1850s, more to provide drainage and remove the offensive presence of sewage than to protect public health, but the result was the same. The tireless work of John Snow in London linked the incidence of cholera to the victims' water source; the simple removal of the pump handle from the contaminated public water well effectively decreased the local incidence of cholera. His colleague Henry Whitehead was initially skeptical of Snow's conclusion but became a believer after he linked the washing of soiled diapers from an infant cholera victim to the same well. The wash water drained to a basement cesspool that was near the well. Upon inspection of the cesspool, a faulty connection to the sewer was found restricting the flow to the sewer and backing sewage into the well. No microscopic or bacterial analysis, simply a linking of cause and effect.

Similar situations could have and probably did exist in the city, but with poor drainage it didn't take an epidemic to teach that getting the water off the land would improve public wellbeing.

Building Sewers in Chicago, 1857–1888

The land surface in the small city was only six to ten feet above the 1847 low water level in Lake Michigan; that level became the Chicago City Datum (CCD). But the lake was not at its low level often, and riparian areas along the rivers were frequently inundated or threatened with inundation. Away from the rivers the land drained slowly, especially when the level of the lake rose with a strong north wind. Further restrictions to drainage were the construction of dock walls and placing fill in low riparian areas to facilitate the mooring of boats. Since the City had no control over water levels in the lake, the only option would have to be control of the land.

The first City ordinance requiring raising the streets and listing elevations was adopted in March 1855; it applied to the area north of Madison Street from the lake to the South Branch. The first placement of fill occurred at the northeast corner of Clark and Lake Streets under the supervision of S.S. Greeley, city surveyor. Greeley became instantly unpopular and was subjected to protests by local landowners. Successive ordinances for raising street grades were adopted in 1857 and 1871, eventually applying to all low-level areas of the city. Raising the streets had to occur for the city to be sustainable.

Under statutory authority passed in Springfield early in 1855, the City leaders appointed the first Board of Sewerage Commissioners and brought in Boston's Ellis S. Chesbrough to be the commission's chief engineer. Benefitting from the experience of engineers in other cities such as London, Chesbrough gathered information and prepared a plan for about seven of the eighteen square miles then comprising the city. The region to be sewered was bounded by Division Street, Lake Michigan, Sixteenth Street, and Ashland Avenue. The commission invited public comment on the plan in December 1855 and sewer construction began in 1856.

Ward boundaries were based on population and political considerations, and rarely followed the river channels. However, the channels were essential for drainage and Chesbrough used the Chicago River main stem and its two branches to divide the city into three districts. In each district, sewers generally drained to the nearest channel or to the lake. After sewer construction began, Chesbrough was sent on a fact finding tour of European cities to learn state-of-the-art sewerage practices. The tour provided Chesbrough with practical and beneficial knowledge; it also showed him that the City's plans for constructing sewers were comparable to plans in Europe's leading cities.

In the interest of economy, shallow sewers were built: essentially, sewers were set on the ground surface or in a shallow trench, then fill was added over the sewer to raise the street to the level specified in the street elevation ordinance. Raising the street provided for sufficient sewer gradient for the flow of sewage to be self-cleaning over the length of the sewer. The sewers were sized to accommodate rain intensity of one inch per hour. That drainage was sufficient in wet weather until the areas draining through each sewer became more developed with larger impervious areas of pavements and roofs.

Chesbrough also considered deep sewers and pumping for drainage. In the end, though, he favored shallow sewers. Pumping would have required coal-fired, steam-driven pumping stations at several locations and the City would have difficulty supporting the initial and ongoing costs. More importantly, the perennial marshy condition of the city needed to be remedied, and raising the grade of the streets would have a long-term benefit. Deep sewers and pumping stations were used later for parts of the city that were topographically low or distant from a channel or the lake.

Most sewers reached the river with less than a mile of length. The longest, about 2.5 miles, drained from Western Avenue eastward to the South Branch. Initially, sewers were sized according to formulae used in London, which Chesbrough brought back from his inspection tour. However, after observing the performance of sewers in the city for 20 years, the formulae were revised to adapt to Chicago's stormier local weather.

The city was growing rapidly with industrial development and shipping via the Great Lakes, Illinois & Michigan Canal, and railroads. The city provided crucial support and industrial activity for the Union during the Civil War. It raised a militia and sited military facilities, including Camp Douglas immediately south of the city. By the end of the Civil War, only nine years after the first sewers were built, the city was bounded generally by Fullerton Avenue on the north, the lakefront on the east, Thirty-Ninth Street on the south, and Western Avenue on the west; Camp Douglas was now within the city. By 1871, before the Chicago Fire in October, the city had added about 11 square miles in a large annexation on the West Side. The fire recovery had a chilling effect on territorial expansion and there were no further annexations until 1887. Fullerton Avenue and Thirty-Ninth Street were the north and south city boundaries, respectively. The west boundary, Crawford Avenue, was later renamed Pulaski Avenue.

Sewer installation in the annexed areas was sporadic. The new areas were sparsely developed and plans called for widely spaced sewers of lesser size than in the downtown area; it was believed these areas would not become as densely developed as the city center. After flooding was experienced, and development was faster and denser than anticipated, additional sewers of larger size were installed. Land along the lakefront was generally lower in elevation and poorly drained, and stormy weather off the lake exacerbated the poor drainage. Dumping of waste materials was prevalent on both North and South Sides, particularly from the Illinois Central Railroad which ran along the lakefront on the city's South Side. The siting of the

Civil War's Camp Douglas at Thirty-Third Street and Cottage Grove Avenue had been a political decision in spite of military doctors' thoughts that the site was unsuitable due to its poor drainage. Sewers discharging to the lake were eventually installed at the camp.

Descriptions of sewers typically refer to *branch*, *relief*, and *trunk* sewers. A branch sewer receives drainage only from adjoining properties, normally within a block or two, and is smaller in size and shorter in length; it discharges into a trunk sewer. A trunk sewer receives discharge from branch sewers and drainage from the properties adjacent to the trunk sewer; it is larger in size and longer. It discharges to an intercepting sewer or receiving waterway. A relief sewer, as the name implies, is built to supplement the capacity of existing sewers, relieving them during overload conditions.

Villages on the periphery of the city were also building sewers, most of which discharged to the nearest river. However, sewers in Evanston and other North Shore suburbs discharged to the lake. Before being annexed into the city, the villages of Hyde Park and Lake View were building sewers that discharged to the lake. More about Evanston, Lake View, and other north suburbs is found in subsequent chapters. Hyde Park and other south suburbs will be discussed in later books.

North District

The Chesbrough plan initially included three trunk sewers in Clark, Franklin, and Rush Streets draining south to the Chicago River. By annexation, the north boundary of the city moved from North Avenue to Fullerton Avenue before the sewer plan was adopted and the sewer in Clark Street extended to Fullerton. As development of the area progressed, trunk sewers were added in other north-south streets. East-west trunk sewers discharging to the North Branch were installed in Chicago and North Avenues. Another sewer in Division Street discharged into a north-south trunk sewer in Halsted Street, which discharged to the North Branch Canal. A continuous branch sewer in Kinzie Street connected many north-south trunk sewers. The added area west of Lincoln Avenue was served by trunk sewers in Center Street, later renamed Armitage Avenue, and in Fullerton Avenue and Webster Avenue, discharging westward to the North Branch. The Center Street trunk sewer discharged to the North Branch via Clybourn Place, later renamed Cortland Street. (See Figure 2.)

Until 1889 there were no further annexations that affected the North District, which was by far the smallest of the three districts. The Fullerton Avenue trunk sewer was alone in the street until the Fullerton Avenue Flushing Tunnel was built in the late 1870s. (See Appendix A-3.)

West District

Trunk sewers were initially installed in alternate east-west thoroughfares and eventually in all east-west through streets from North Avenue to Harrison Street, many extending west to the city boundary at Western Avenue. A continuous trunk sewer in Milwaukee Avenue connected several east-west trunk sewers, discharging to the Kinzie Street trunk sewer. South of Harrison Street, rather than having the east-west trunk sewers extend as far west as Western Avenue, they reached only as far as the point where it would be the same distance for a north-south trunk sewer to reach the South Branch or a trunk sewer in Twenty-Second Street discharging to the West Fork through a trunk sewer in Paulina Street. Although there were many navigation slips to the north of the South Branch and West Fork, these slips were between street rights-of-way and sewers didn't discharge to these slips. The upper end of the trunk sewer in Ashland Avenue connected to the trunk sewer in Harrison Street. The same applied to sewers in Center Avenue, later renamed Racine Avenue, and Twelfth Street. (See Figure 2.)

Following the Civil War, the city annexed a large area to the west bounded by North Avenue on the north, the Illinois & Michigan Canal on the south, and Crawford Avenue on the west. Installing sewers took several years due to the pace of development and recovery efforts following the Chicago Fire in 1871. Eventually, the area south of Kinzie Street was served by trunk sewers in north-south thoroughfares, each half mile draining south to the Twenty-Second Street trunk sewer. Two trunk sewers in Leavitt Street and Western Avenue relieved the Twenty-Second Street sewer, discharging to the West Fork. Since the area south of Twenty-Second Street was mostly industrial, only two public sewers were installed in Kedzie and Lawndale Avenues; they drained to the West Fork.

In the newly annexed area north of Kinzie Street, a trunk sewer in Ohio Street flowed east from Crawford Avenue to Kedzie Avenue, then north to Augusta Boulevard. Near this last point, another Grand Avenue trunk sewer flowing from Crawford Avenue turned east in Augusta to California Avenue then north to North Avenue. The trunk sewer in North Avenue flowed east from Crawford Avenue to the North Branch. A trunk sewer in Division Street connected to the California Avenue trunk sewer and flowed east to the North Branch. In all, the West District was the largest sewer district until the annexation of 1889.

South District

A trunk sewer installed in Michigan Avenue from Thirty-First Street north to the Chicago River had one outlet at the river and two other outlets discharging to the lake at Twelfth and Twenty-Second Streets. East of State Street, branch sewers flowed eastward into the Michigan Avenue trunk sewer. West of State Street and north of Randolph Street, short trunk sewers were installed in each north-south street discharging to the Chicago River, with branch sewers in the few east-west streets. West of State Street and south of Randolph Street to Twelfth Street, trunk sewers were installed in the east-west through streets discharging to the South Branch, and branch sewers were installed in the north-south streets.

South of Twelfth Street the trunk sewer layout is complicated, as shown in Figure 2, and most networks of branch and trunk sewers ended up discharging to the South Branch or South Fork at several locations. Until the Civil War, the south boundary of the city was Thirty-First Street; during the war, the south boundary was extended one mile to Thirty-Ninth Street.

In addition to the two outlets to the lake at Twelfth and Twenty-Second Streets, another was located at Thirty-Fifth Street. It served a sewer network for the area bounded on the north by Thirty-First Street, on the east by the lakefront, on the south by Thirty-Ninth Street, and on the west by Wentworth Avenue. Camp Douglas was closed at the end of the Civil War; in December 1865, with the buildings razed and the availability of sewers, the area was ripe for development.

Prior to annexations in 1889, only three sewers discharged to the lake from the city, all in the South District: Twelfth Street, Twenty-Second Street, and Thirty-Fifth Street. With the 1889 annexations, the South District became the largest district.

Illinois & Michigan Canal

The South Branch and West Fork formed the boundary between the South and West Districts. Just to the south of the West Fork was the I&M Canal, which began in Bridgeport on the east side of Ashland Avenue at the confluence of the South and West Forks, the present location of Origins Park. When the canal opened in 1848, it was used for navigation, the purpose for which it was built. But ten years later, with sewer construction, the canal was also used for relief of the offensive conditions in the South Branch and two forks. The water level in the canal at Bridgeport was approximately eight feet above CCD and a pumping plant was built to lift water into the canal from the South Branch during dry weather. Sewage-laden river water could be discharged by gravity into the canal during and following wet weather periods. Since dry weather prevailed most of the time, continuous pumping was necessary; also, the canal had limited discharge capacity and excessive pumping could result in the canal overflowing its banks. (See Figure 3.)

By 1865 the condition of the South Branch was intolerable. Chesbrough, along with other leading engineers including William Gooding, Illinois & Michigan Canal chief engineer during canal construction, recommended to the City that the deep cut plan for the Illinois & Michigan Canal be implemented. This original plan for the canal had not been implemented due to financial constraints during the construction period from 1836 to 1848. The deep cut would provide the required capacity to discharge sewage-laden river water during dry weather. The City adopted the plan and obtained permission from the canal commissioners to build the deep cut.

The original pumping station was removed from service when construction started in 1867; because of contractor problems, the deep cut wasn't completed until 1871. After a few years, the deeper canal proved ineffective because of sediment deposition, the sediment coming from increased sewage and erosion of steep bluffs in the area southwest of Willow Springs. A new pumping station, built and placed in service in 1883, continued to serve until the Sanitary & Ship Canal was opened in 1900. The pumping station capacity was 833 cfs, but it rarely was sufficient to draw water from the lake; it mostly drew offensive water from the South Fork and South Branch.

Once the Sanitary & Ship Canal was opened on January 17, 1900, the I&M Canal was relieved of service to improve river conditions and the pumping station served only to supply water for navigation. The City ran the pumps continuously through mid-July 1900, at which time the station was turned over to the District. Since the opening of the Sanitary & Ship Canal lowered the water level in the South Branch by several feet, the District felt an obligation to supply water for navigation.

City Annexations in 1889

As described in *Building the Canal to Save Chicago*, the storm and flood of August 1885 led to the creation of the District in 1889 and construction of the Sanitary & Ship Canal. The canal had a significant impact on drainage and public health in the city after it was opened in January 1900. However, 1889 was notable for another reason; remarkably, the city nearly quadrupled in area in one year by adding to the size of all three sewer districts. The total city area went from 47.3 square miles to 174.1 square miles. The growth in population was not as dramatic—the new areas, less densely populated, added only 225,000 people. City population more than doubled between the 1880 and 1890 censuses, from 503,185 to 1,099,850, but the annexed population was less than half of the nearly 600,000 increase.

The annexed area was largely vacant, poorly drained, with scattered settlements. Towns were growing on the periphery of the area and the city needed more territory for its growing population. The 1890 census showed significant population in the expanded South Side, which was south of Thirty-Ninth Street and east of Halsted Street, near the mouth of the Calumet River, and in Pullman and West Pullman. In the less populated North Side, population was centered near Ashland and Belmont Avenues.

Typically, small sewers of limited extent served most communities; without proximity to a river, these sewers discharged to a small creek or drainage ditch. City engineers struggled to cope with the additional infrastructure. Communities that grew rapidly soon overwhelmed the drainage systems which had been designed and built for smaller populations.

The South District was the most impacted by the 1889 annexation. East of State Street, the south boundary was moved more than 12 miles from Thirty-Ninth to 138th Street with the addition of the Township and Village of Hyde Park, adding approximately 48 square miles. West of Halsted Street, the south boundary was moved six miles to Eighty-Seventh Street with the addition of the Town of Lake, adding approximately 36 square miles. (See Figure 4.)

In addition to the sewer systems in these areas, the City took on several pumping stations and numerous sewer outfalls discharging to Lake Michigan. The pumping station names and locations follow:

- Seventieth Street Pumping Station, Seventieth Street and Yates Boulevard
- Seventy-Third Street Pumping Station, Seventy-Third Street and Railroad Avenue, later renamed Exchange Avenue
- Sixty-Ninth Street Pumping Station, Peoria and Sixty-Ninth Streets
- Woodlawn Pumping Station, Sixty-Second Place and Illinois Central Railroad, today the Metra Electric Line

Three of the above pumping stations discharged to Lake Michigan. The Sixty-Ninth Street Pumping Station discharged to the Halsted Street sewer, which flowed north to Thirty-Ninth Street and discharged to the East Arm of the South Fork, aka the Stock Yards Slip.

The West District added a majority of the Town of Jefferson, approximately 22 square miles, and moved the west boundary to Austin Avenue north of Irving Park Boulevard and to Harlem Avenue between Irving Park Road and North Avenue. South of North Avenue, the west boundary was moved to different streets near Cicero Avenue, adding approximately six square miles. Numerous small sewers were included for scattered settlements, mostly discharging to the North Branch or the West Fork of the South Branch.

The North District was the least expanded, adding approximately 15 square miles. The north boundary was moved five miles from Fullerton Avenue to Devon Avenue; that region included, the remainder of the Town of Jefferson and the City of Lake View. Existing sewer systems in the annexed area to the North District discharged to the North Branch in the City of Lake View. Farther north, in Rogers Park and West Ridge, about three square miles were annexed in 1893, moving the north city boundary another 1.5 miles from Devon Avenue to Howard Street and the south line of Calvary Cemetery. Sewers in Rogers Park drained to the lake; the ground was low, so pumping was necessary. Three systems used equipment called Shone ejectors to lift collected sewage into a gravity sewer. Ejectors use compressed air to force liquid out of a tank to a higher level. Sewage and drainage from neighborhood sewers fill an underground tank by gravity. Once full, compressed air is let loose at the top of the underground tank. The air pressure forces sewage out of the tank, lifting it to a higher sewer, where it is discharged to the lake by gravity. Two ejectors each were located at Albion and Glenwood Avenues, at Ashland and Morse Avenues, and at Ashland and Touhy Avenues. By 1907, the practice was eliminated by construction of the lakefront intercepting sewer. That construction is described in the next chapter.

By the turn of the century, the populated area expanded south to Seventy-Ninth Street, west to Western Avenue south of the West Fork, west to Cicero Avenue north of the West Fork, and north to Lawrence Avenue. North of Lawrence Avenue, the population was centered in Andersonville and Rogers Park along Clark Street.

Pure Water Commission

While the District was busy constructing the Sanitary & Ship Canal and the City was building sewers to accommodate the added territories and growing population, some leaders were thinking of the future and how the City and District could work together in resolving pollution of Lake Michigan. In September 1896, Mayor Swift stopped the construction of a new sewer that was to discharge to the lake and appointed the Pure Water Commission to develop a plan to eliminate existing discharges and define responsibilities in implementing the plan. In February 1897 the commission reported a plan: construct intercepting sewers along the lakefront, one for the North Side and one for the South Side; collect all sewage discharging to the lake; and construct two pumping stations, again, one for each side of the city, to pump the sewage to the nearest river channel. At the urging of the District, the plan included using lake water at each pumping station to dilute the sewage so that it would be less offensive when reaching the river channel. The commission also suggested the City and District work together to implement the plan and share the costs.

After approval by the city council, the department of public works began execution of the plan in March 1897 for both North and South Side intercepting sewer systems. Later conferences between the City and the District resulted in agreement in January 1900 for the inclusion of lake water pumps, cost sharing, construction by the City, and operation and maintenance by the District. The details of the North Side infrastructure are described in the next chapter; the details of the South Side infrastructure will be described in later books.

Sewersheds

The Pure Water Commission plan and the cooperation between the City and District changed how the land was divided into sewersheds or drainage areas. As mentioned above, Chesbrough originally divided the city into three districts bounded by the river channels. The plan for lakefront intercepting sewers and pumping stations began to define new sewershed boundaries. Later, when the District began building intercepting sewers and sewage treatment plants, new sewersheds were defined to align with sewage treatment plants, not the nearest channel.

Fullerton Avenue had a lasting effect on drainage as it became the approximate boundary between sewage and some stormwater that drained northward to the North Side Sewage Treatment Works on Howard Street, completed in 1928. This plant was renamed the O'Brien Water Reclamation Plant in 2012. South of Fullerton Avenue, sewage and some stormwater drained southward and westward to the West Side Sewage Treatment Works, which opened in Stickney in 1930. This plant is presently part of the Stickney Water

Reclamation Plant on West Pershing Road. A third sewershed drained to the Calumet Sewage Treatment Works on 130th Street. Opened in 1922, it is presently named the Calumet Water Reclamation Plant. The Stickney and Calumet sewersheds are discussed in later books.

A term that appears above, *sewage and some stormwater*, is used because the sewers in the city and in the older surrounding suburbs were designed to use the same pipe to convey both sewage and stormwater. During dry weather, only sewage was conveyed, and in wet weather, both sewage and stormwater were conveyed, with stormwater most often being the major component. The District intercepting sewers collected sewage and some stormwater from municipal combined sewer systems and conveyed it to a treatment or water reclamation plant. The intercepting sewer would accept sewage and some stormwater up to its limiting capacity before the excess flow was discharged to the canal system or a local river. The intercepting sewer's ability to accept stormwater varied depending on time of day, day of the week, and the intensity, distribution, and duration of rainfall.

The City continued to build sewers; by 1900 it had installed a total of 1,481 miles, of which nine miles were replacement of smaller with larger sewers. By 1905, the accumulated outlet discharge capacity of all sewers reached approximately 10,000 cfs, the same as the design capacity of the Sanitary & Ship Canal. By 1925, the aggregate City sewer outlet capacity reached 20,000 cfs, which is the practical maximum discharge capacity of the Sanitary & Ship Canal between Sag Junction and the Lockport Controlling Works. (See Figure 5.) The doubling of City sewer outlet capacity in 20 years illustrates how rapidly the City was providing drainage capacity for its land area.

Incidentally, it is not intended to suggest—and it does not mean—that the canal system today is overloaded. The canal system has considerable storage capacity to accept and attenuate peak overflows from sewers. In addition, the District's Deep Tunnel system, including reservoirs, provides additional storage and conveyance capacity for excess stormwater.

Over time, the City has built additional relief sewers not only to address continued development of the land surface, but also to address local drainage insufficiency, changing design criteria and cultural practices. The distribution, frequency, and intensity of storms have changed over time resulting in changing design criteria. Many changing cultural practices result in the sewers having to convey more stormwater runoff in a shorter period of time. These cultural practices include covering more land surface with impermeable pavements and roofs, replacing smaller buildings with larger buildings, more vehicles requiring more parking lots, landscaping with less permeable surfaces, etc.

References

Brown, George P. *Drainage Channel and Waterway: A History of the Effort to Secure an Effective and Harmless Method for the Disposal of the Sewage of the City of Chicago, and to Create a Navigable Channel Between Lake Michigan and the Mississippi River*. Chicago: R.R. Donnelley & Sons Company, 1894.

City of Chicago. *The Chicago Sewer System, 1856–1956*. Chicago: Department of Water and Sewers, 1956.

City of Chicago. Department of Public Works. Annual Reports, 1889 through 1910.

Cooley, Lyman E. *Lake Diversion at Chicago*. Chicago: Sanitary District of Chicago, February 1913.

Grossman, James R., Ann Durkin Keating, and Janice L. Reiff. *Encyclopedia of Chicago*. Chicago: University of Chicago Press, 2004.

Hill, C.D. "The Sewerage System of Chicago." *Journal of the Western Society of Engineers* 16, no. 7 (September 1911): 545.

Johnson, Steven. *The Ghost Map*. New York City: Riverhead Books, Penguin Group, 2006.

Keller, David L. *The Story of Camp Douglas*. Charleston, SC: The History Press, 2015.

Mayer, Harold M. and Richard C. Wade. *Chicago, Growth of a Metropolis*. Chicago: University of Chicago Press, 1969.

Trantowski, Elizabeth, Howard Rosen, and Lawrence E. Lux. *A History of Public Works in Metropolitan Chicago*. Kansas City, MO: Chicago Metro Chapter, American Public Works Association, 2008.

wikipedia.org.

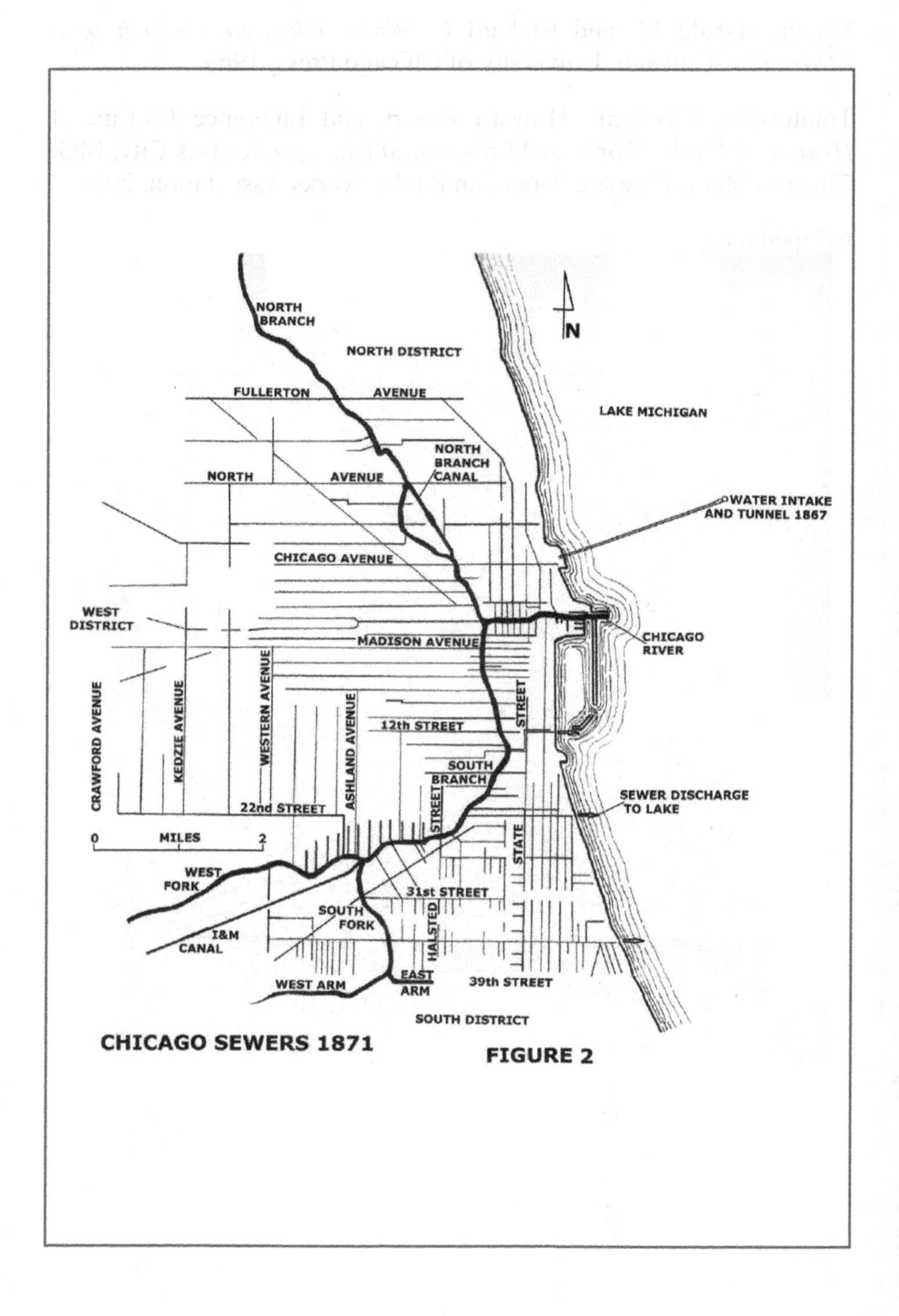

CHICAGO SEWERS 1871

FIGURE 2

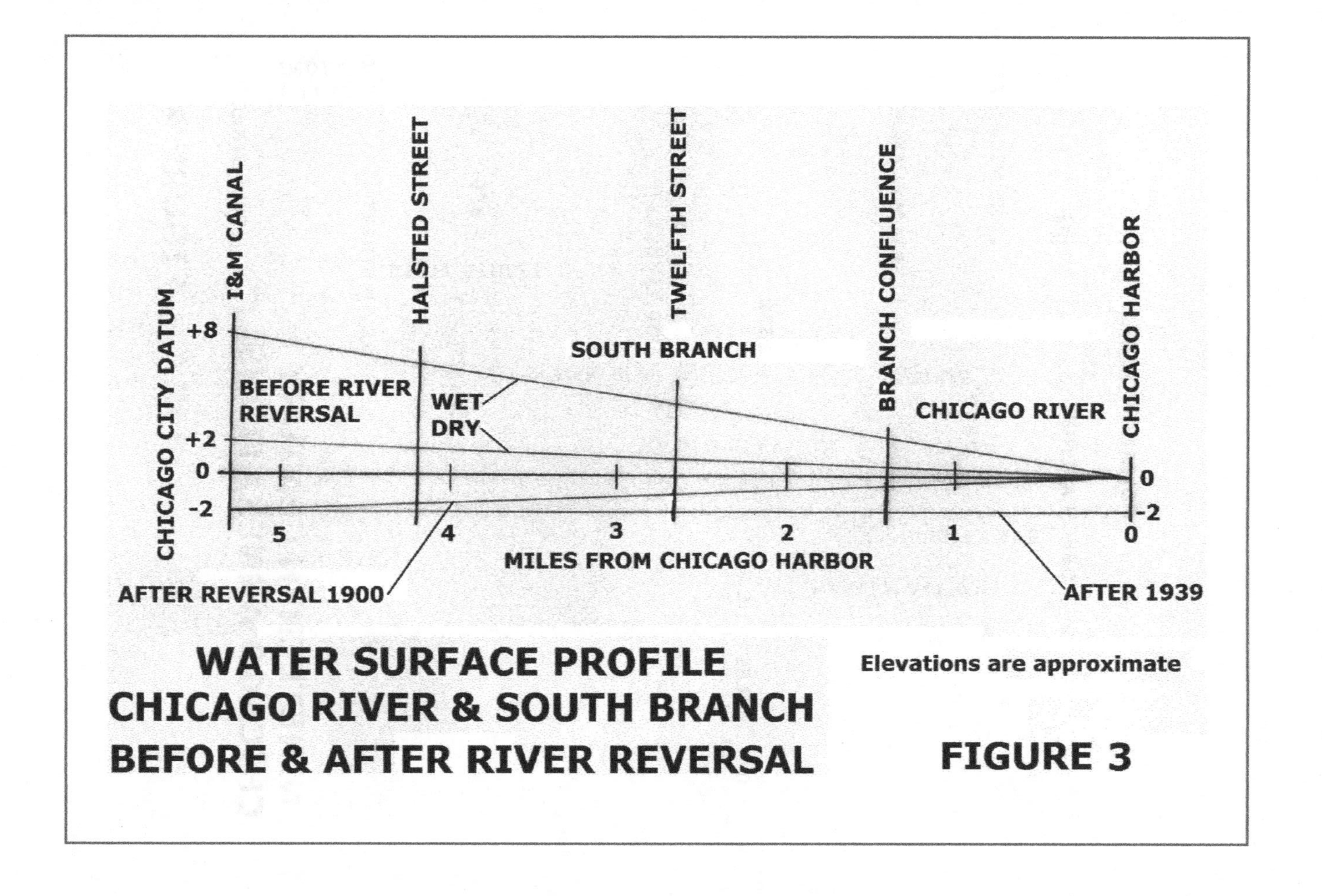

WATER SURFACE PROFILE
CHICAGO RIVER & SOUTH BRANCH
BEFORE & AFTER RIVER REVERSAL

FIGURE 3

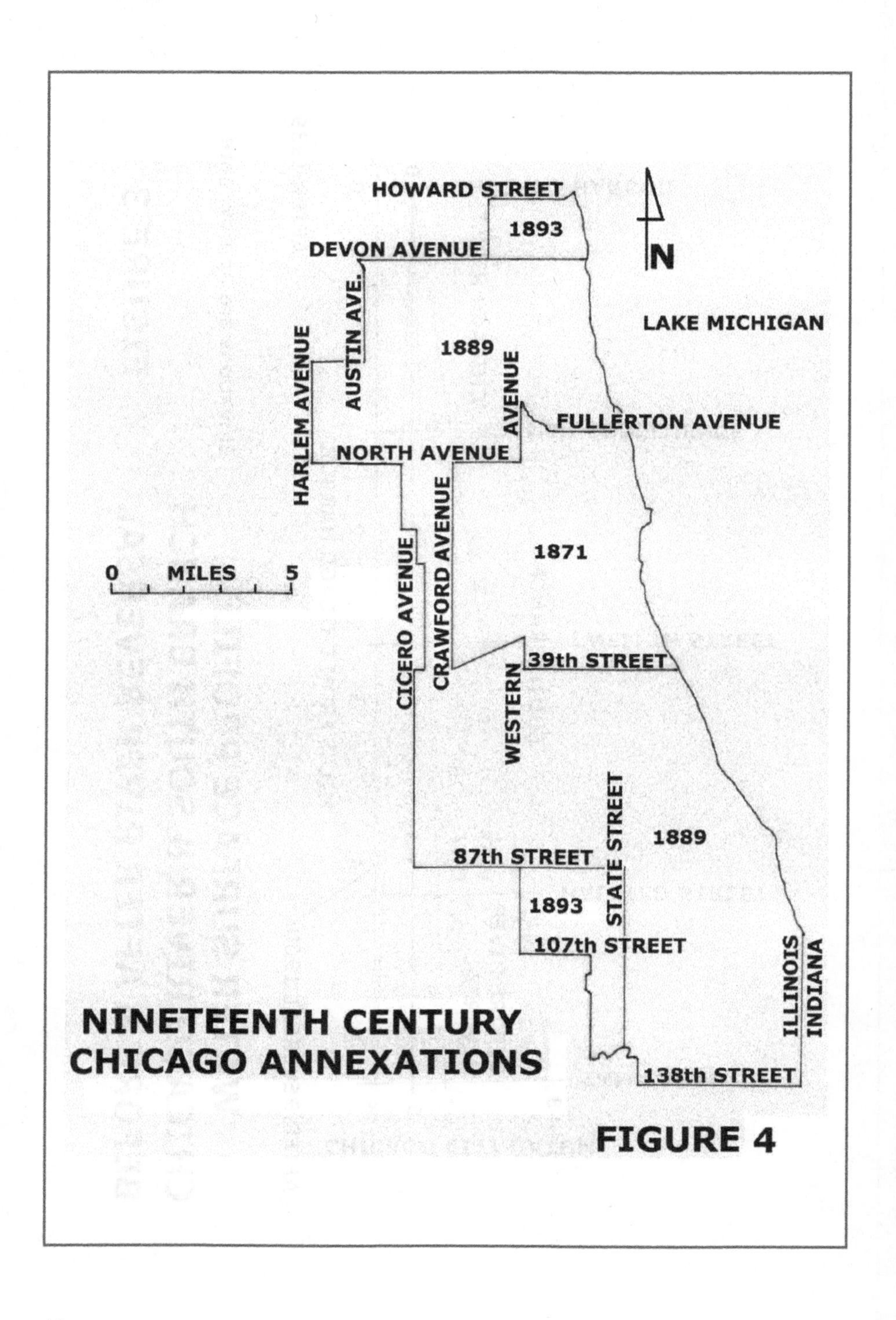
HOWARD STREET
1893
DEVON AVENUE
N
LAKE MICHIGAN
AUSTIN AVE.
HARLEM AVENUE
AUSTIN AVENUE
1889
AVENUE
FULLERTON AVENUE
NORTH AVENUE
CRAWFORD AVENUE
CICERO AVENUE
1871
0 MILES 5
39th STREET
WESTERN
STATE STREET
1889
87th STREET
1893
107th STREET
ILLINOIS
INDIANA
138th STREET
NINETEENTH CENTURY
CHICAGO ANNEXATIONS
FIGURE 4

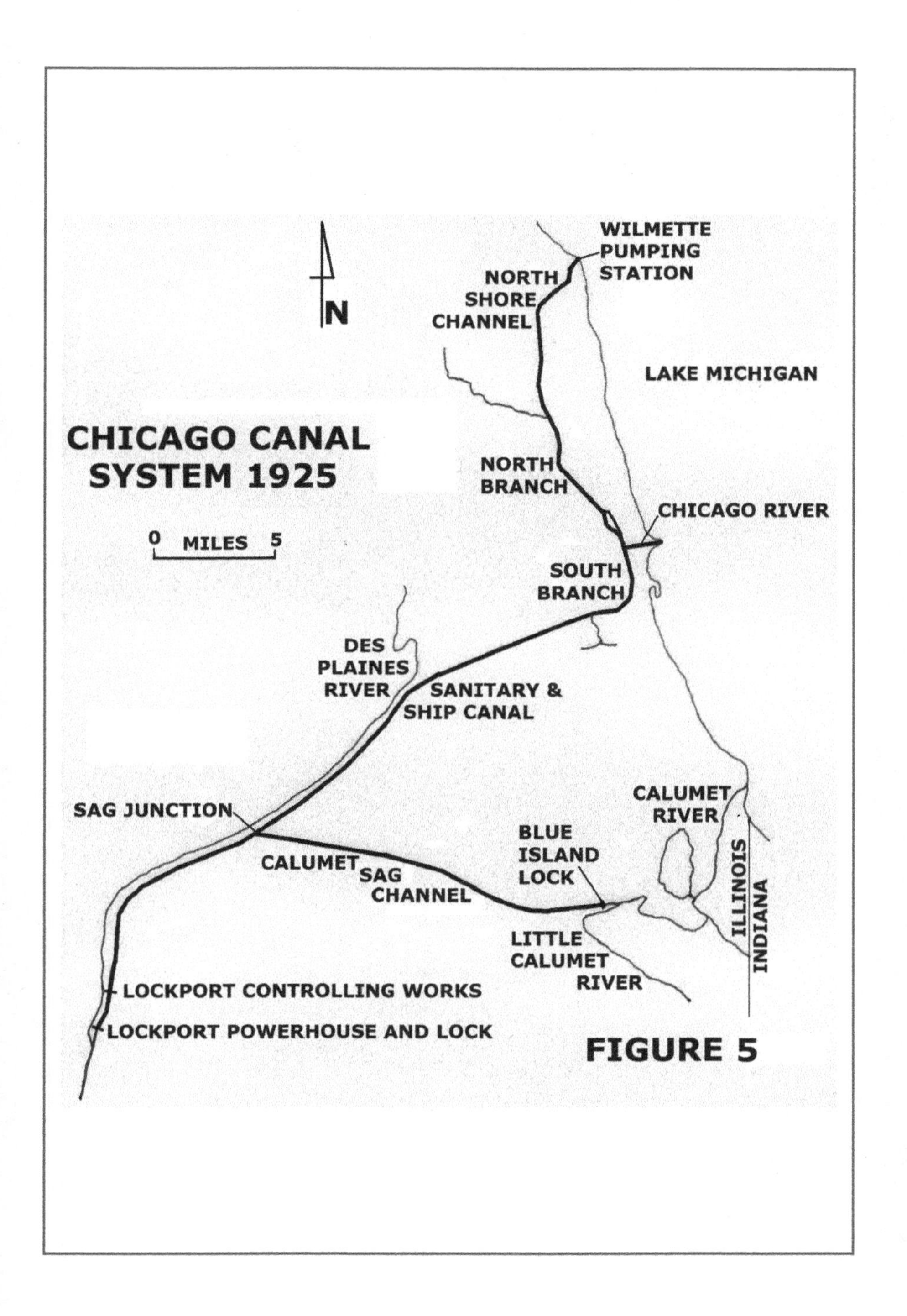
N
WILMETTE
PUMPING
STATION
NORTH
SHORE
CHANNEL
LAKE MICHIGAN
CHICAGO CANAL
SYSTEM 1925
NORTH
BRANCH
CHICAGO RIVER
0 MILES 5
SOUTH
BRANCH
DES
PLAINES
RIVER
SANITARY &
SHIP CANAL
SAG JUNCTION
CALUMET
RIVER
BLUE
ISLAND
LOCK
CALUMET
SAG
CHANNEL
ILLINOIS
INDIANA
LITTLE
CALUMET
RIVER
LOCKPORT CONTROLLING WORKS
LOCKPORT POWERHOUSE AND LOCK

FIGURE 5

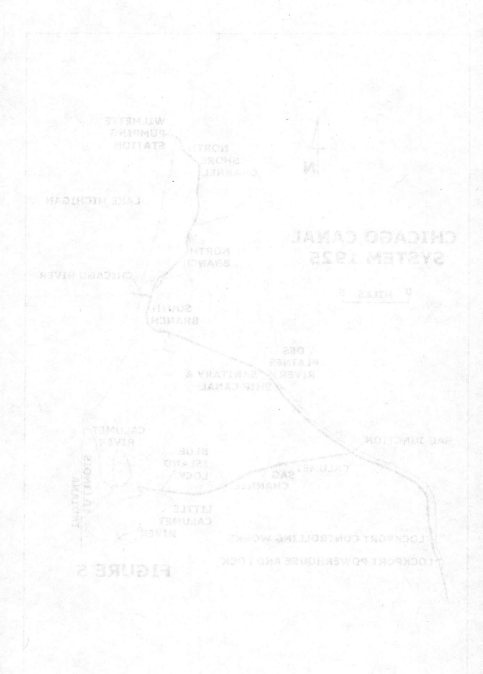

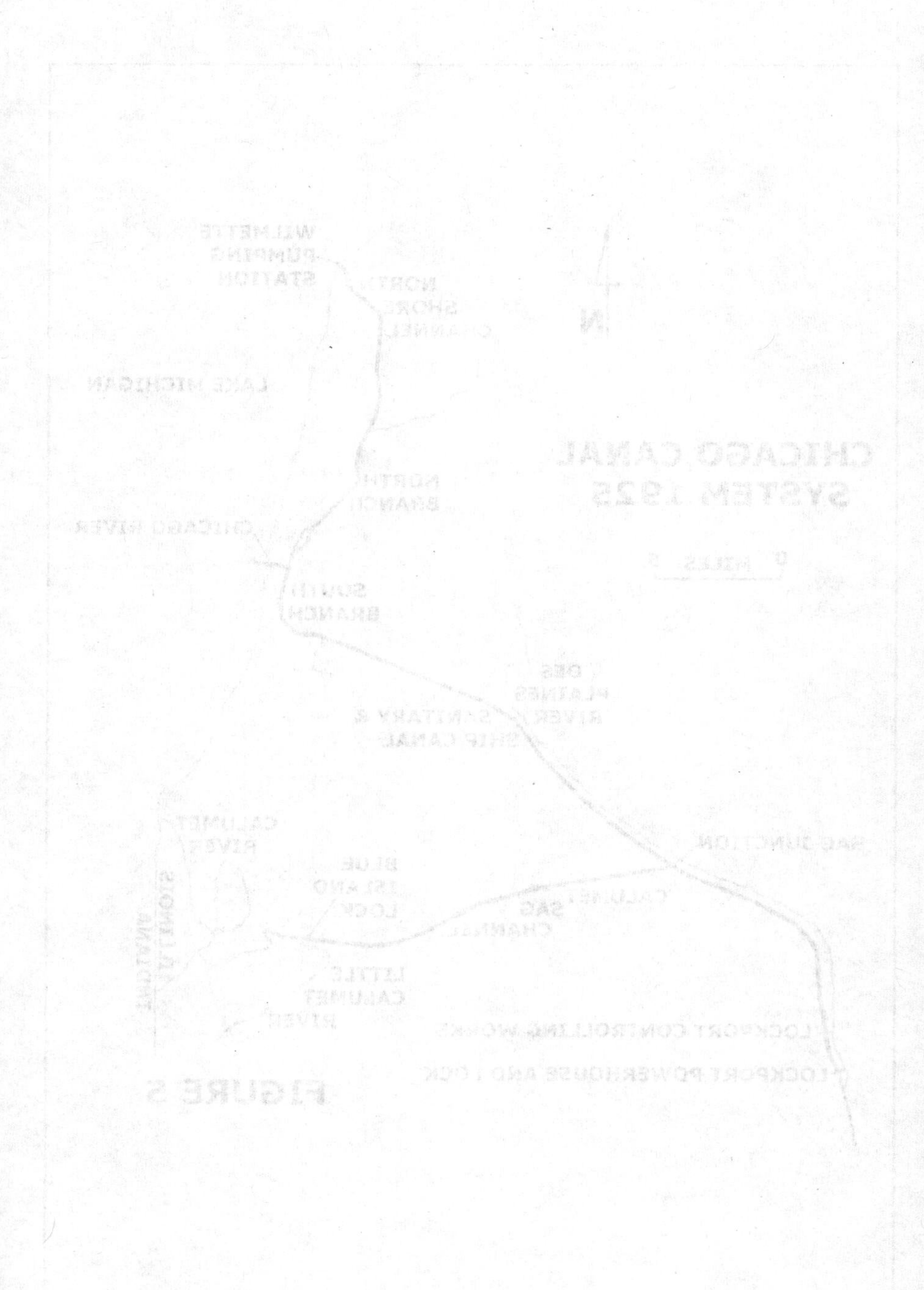

Chapter 2

New Digs for the North Branch

Introduction

South of Belmont Avenue, the North Branch was becoming grossly polluted from trade wastes and sewage during frequent periods of low flow. Occasional floods from the watershed flushed the pollution, but also caused damage to riparian property and structures. The condition of the North Branch was of as much concern for as long a time as the South Branch. However, the opportunities for corrective action on the North Branch were more difficult to achieve. As population grew along the lakefront, increasing amounts of sewage threatened the water supply taken in at the existing off-shore cribs east of Chicago Avenue; the increased sewage also threatened a planned crib east of Wilson Avenue. Cooperation between the City and District was essential to carry out a plan for the North Branch and North Side of the city.

Farther north, beyond the jurisdiction of the District, growth in the North Shore suburbs and sewage discharge to the lake were beginning to threaten local water supplies. In 1889, Evanston and other North Shore communities had opted out of the formation of the District, but interest grew once it was obvious how reversing the flow in the Chicago River and South Branch had reduced mortality and made the city more livable. It was now evident that expanding the District would be necessary to protect public health and Lake Michigan. (See Appendix A-1.)

North Branch

In 1903, the General Assembly expanded the corporate jurisdiction of the District to the Cook-Lake County line. That act enabled the District to address the problem of sewage from the north lakefront suburbs in Cook County. A channel was needed for two purposes: to serve as an outlet for sewers discharging to the lake, and as a means of diverting lake water for dilution. There was no question where the channel would be, but before such a channel could be constructed the North Branch needed to be improved; it stood as a barrier to effective drainage north of Belmont Avenue. (See Figure 6.)

The Chicago River and Harbor served as a port of refuge for military vessels; as such, they were critical to the defense of Fort Dearborn. Since the early years of the city, the Corps of Engineers (Corps) had been committed to maintain the Chicago River and Harbor. In 1894, following congressional authorization, the North Branch was dredged to provide a 16-foot depth as far north as Belmont Avenue. This dredging maintained adequate depths for navigation, but only in a few places was the channel widened to remove obstructions. The Corps expressed frustration at how rapidly their work could be reversed by sedimentation from sewage and industrial waste discharged to the river. In 1904, the Corps built the turning basin on the North Branch at North Avenue.

Land along the North Branch north of Belmont Avenue was less intensely developed. Riparian property was occupied by clay pits and brick manufacturing plants, recreational areas, and small farms. Most riparian property was low and subject to frequent flooding. The route of the stream channel, generally north to south, meandered excessively and was shallow and slow moving. North of Lawrence Avenue the route of the stream channel changed, coursing from northwest to southeast. Thus the channel moved farther west, away from Lake Michigan. (See Figure 6.)

Eliminating Sewage Discharges to the Lake

While the District was designing an improved channel for the North Branch, the City department of public works took the lead in the design and construction of the Pure Water Commission plan facilities explained in the prior chapter. Two intercepting sewers were constructed. One was from Chase Street, later extended to the north city limit, and flowed south in Sheridan Road to Lawrence Avenue; a second, from Surf Street, later extended to Fullerton Avenue, flowed north in Lakeview Avenue, Sheridan Road, Irving Park Road, and Clarendon Avenue to Lawrence Avenue. The two intercepting sewers joined in Lawrence Avenue and flowed west to the pumping station. Upon completion, City sewers discharging to the lake were connected to the intercepting sewers and the lake outfalls were sealed. (See Figure 7.)

For lake water, a separate intake crib, intake tunnel, and intake conduit were constructed to the pumping station from 1,445 feet offshore. Separate pumps were used for lake water so the dilution rate could be controlled. (See Figure 8.)

In 1898, contracts were awarded by the City for construction of the intercepting sewers and the work was completed the following year. Contracts for construction of the Lawrence Avenue conduit were awarded in 1900, but dissatisfaction with one contractor's progress soon became an issue. The contract was forfeited, the City resumed construction using day labor, and the contractor sued. The trial court found in favor of the contractor and ordered the City to cease construction and re-advertise the work. The City obeyed, and in 1905 construction of the conduit resumed; it was finally completed by 1907. Until construction of the Lawrence Avenue Pumping Station and Lawrence Avenue conduit was completed, a temporary pumping plant was installed so sewage collected in the intercepting sewers could be discharged to the lake.

In November 1901, the City initially sited a permanent pumping station near the North Branch and requested District approval of the site and conduit. The City was also considering a site for a pumping station near the intersection of Lawrence and California Avenues. A year later the District suggested the City change the site of the pumping station to conform to the relocation of the North Branch, at that time one block farther west. The delay in District response was due to the planning for the new channel for the North Branch and North Shore Channel.

In February 1903, the City notified the District of its interest in relocating the site of the pumping station from near the North Branch to near the lakefront; the District's agreement was needed so the change could be submitted to the city council for approval. The new site, purchased in 1904, was located on the north side of Lawrence Avenue between Evanston Avenue, later renamed Broadway, and the Chicago, Milwaukee & St. Paul Railroad. The railroad ran on grade and, beginning in 1908, the tracks were used by the Northwestern Elevated Railroad. Still later those tracks were elevated; today it is the route of the Chicago Transit Authority Red Line.

The new location for the pumping station lowered the cost of the conduit as it could be shallow rather than deep, thus lessening the depth and cost of excavation. The pumping station was essentially two stations in one, the sewage and lake water combining after pumping and then flowing west by gravity in the conduit to the North Branch. The City proceeded to advertise the construction contracts despite wanting to construct these facilities using its own employees. Upon learning of the City's intent, the contractors again sued and won. In December 1904, the Illinois Supreme Court confirmed the decisions of the trial and appeals courts to require competitive bidding.

To recommend appropriate pumps, the City hired William S. MacHarg, a consulting engineer. In February 1905, his recommendation was forwarded to the District for approval, which was given. The agreement provided for a total pumping capacity of 585 cfs, but Mr. MacHarg recommended pumps that exceeded this capacity in order to allow a comfortable margin for maintenance down time. His recommendations included a centrifugal pump with a capacity of 37.5 cfs for sewage and two centrifugal pumps with regular-use capacity of 125 cfs and 50% greater capacity at reduced head for stormwater. For lake water and standby service, an axial flow screw pump with a capacity of 585 cfs was recommended if other pumps were out of service. Each pump would be driven by a steam engine and the station fitted with coal-fired boilers.

In November 1906, almost a year before the completion of the new North Branch Channel by the District, the City contractors completed the pumping station and conduit and the discharge of sewage and lake water to the North Branch began. Completion of the intercepting sewer and conduit connections required another year.

In November 1910, a contract was executed for the transfer of the Lawrence Avenue Pumping Station, intake, and crib to the District. The conduit was not included in the transfer, but the City indicated that it would be considered at a later time. The conduit was never transferred because the City planned to connect other sewers to it. The agreement provided for the transfer of the intake crib, intake shaft and tunnel, protection piers, pumping station and all contents, and the pumping station property. Due to the ravages of waves and winds over the next 20 years, the intake crib required periodic repair and dredging to maintain its function and capacity.

The City lakefront intercepting sewers drained an area, generally lying east of Clark Street from Fullerton Avenue to the north city boundary, of about five square miles. The pumping station sent sewage and dilution lake water into the Lawrence Avenue conduit, a 16-foot diameter brick sewer, where it traveled 2.2 miles west and discharged to the North Branch. The District made improvements to the pumping station after assuming control, including more efficient boilers. Another improvement, when the City required elevation of the adjoining railroad tracks, was an elevated spur for coal deliveries. This arrangement for drainage, dilution, and sewage lasted for another 20 years until June 1930, when the North Branch Pumping Station at Lawrence and Francisco Avenues was placed in service; at that time the Lawrence Avenue Pumping Station was removed from service. The site reverted to control by the City and the structure remained in place until eventual demolition by the City for reuse of the site.

Channel Improvement Planning

By 1899, the North Branch south of Belmont Avenue had already been deepened by the Corps to a depth of 16 feet. The District determined that north of Belmont an improved channel was necessary to receive the discharge from the Lawrence Avenue conduit and as an outlet for a channel to serve Evanston, Wilmette, and other north suburbs. District engineers had begun field surveys in 1897 for the North Branch and extended those surveys well into Evanston. As soon as the field notes were available, a map was prepared and, before the end of the year, planning began for the improvement of the North Branch north of Belmont Avenue.

South of Lawrence Avenue, the natural channel of the North Branch meandered nearly a half mile to the east, and then back about one-quarter mile to the west north of Montrose Avenue. From Montrose, flowing south to Belmont, the channel meandered east and west less severely. The District decided on a new channel course from Belmont northward to Montrose with slight bends between straight segments to accommodate some of the meandering. However, north of Montrose, the course was straight on a northwest diagonal to Lawrence, completely eliminating the grand meander to the east. (See Figure 9.)

Obviously, the benefits of stream ecology and floodplain storage were not known to, or on the minds of, the City or District leaders when planning this project. These plans were made nearly a decade prior to the adoption of the Burnham Plan and 14 years prior to the creation of the Forest Preserve District of Cook County. The resulting channel improvement reduced 32 acres of floodplain to 14 acres of new channel, shortened 3.2 miles of meandering shallow stream to 2.2 miles of new straight deep channel, and made 18 acres available for development above the floodplain. Despite the losses, the benefit was better drainage with the larger channel and lower water level. The new channel confined to its 180-foot right-of-way does not go out of its banks like the natural channel did.

Through interviews with landowners in the area, the District learned that the right-of-way could be obtained at no cost, an attractive prospect since the tax rate had reverted back to the pre-1895 level. The landowners realized that the new channel would reduce flooding on their land and the spoil from the new excavated channel could be used to fill in low areas on their residual property. In May 1900, the approved plan was to deepen, straighten, and widen the North Branch between Belmont and Lawrence Avenues, provided the adjoining land owners would agree to donate a strip of land 180 feet wide for the new channel. The planned channel was to be excavated to a width of 80 feet and a depth of eight feet below CCD. There would be no spoil piles as the excavated clay would be spread on low areas on the adjoining property.

Land Acquisition

The route was adopted in April 1902, defining the new channel from

Belmont Avenue to Lawrence Avenue and surveys conducted to identify property owners. The course was set for the improvement of the North Branch and negotiations with property owners were progressing so well that contracts were prepared for each land holder. To provide evidence of their good intent to the land holders, City, and the citizens, the District committed to the completion of the construction of the new channel in two years, unless there was a delay due to litigation.

Several land parcels were acquired in 1903 and most were simple transactions; a few required condemnation. Paul O. Stensland owned property within and east of the channel route between Berteau and Belle Plaine Avenues where the natural channel meandered through his property. He was so certain of the enhanced value of his residual property that he agreed to donate $2,000 to the District if the new channel were completed in three years. Due to late completion, the donation was never received. Oliver B. Green and E. Louisa Green owned several lots along the north side of Pensacola Avenue within and east of the channel route. The District agreed to strip the topsoil off of the acquired property and spread it on Green's residual property. Louis S. Owsley's home was on the west bank of the North Branch in the Electric Park subdivision south of Roscoe Street, but within the new channel right-of-way. The District agreed to allow Owsley to stay in his house until construction of the channel required its removal. William Tempel lived west of the North Branch, but within the channel route south of Irving Park Road, and he desired water access from his residual property and was granted access across the land sold to the District. These transactions were all south of Montrose Avenue where there were many property owners.

The Northwest Land Association owned all the property between Montrose and Lawrence Avenues and agreed to let the District have the needed 180-foot wide strip at no cost in exchange for spreading all excavated spoil on low adjoining land, provided the new channel was completed within three years. If the channel was not completed on time, the land would revert to the association. The transaction with the association was approved in November 1903, marking the start of eventual litigation discussed in Appendix A-5.

Channel Excavation and Bridge Construction

The surveyors had the construction work completely staked out by December 1903. While waiting for the start of construction, a permit was issued in February 1904 to the heirs of Joseph Bickerdike to remove approximately 15,000 cubic yards of soil from the channel route, a quantity over and above that required for filling in the old stream bed. The property formerly owned by Bickerdike was located south of Addison Street approximately where the present day Clark Park is located. At the same time, Louis Grimme was granted the right to occupy the land previously sold to the District until such time as needed for construction.

The earlier commitments for completion within two years was changed due to the difficulty and additional time needed to secure the necessary land rights. Design revisions caused by changes in the agreement with the Northwest Land Association also caused a six-month delay to the start of construction. The new completion target was November 1905. Perhaps it was thought that a new channel, a little over two miles in length, wasn't such a big undertaking. Time would tell otherwise.

Construction of the improvement was divided geographically into two contract sections. Section 1 extended from the south line of Belmont Avenue to the north line of Montrose Avenue, a distance of 8,400 feet. Section 2 extended from the north line of Montrose Avenue to the south line of Lawrence Avenue, a distance of 3,100 feet. Section 1 was intended to be excavated by dredging because of frequent crossing of the river channel, while Section 2 was intended to be excavated in the dry since it was mostly distant from the meandering river channel. In February 1904 two contracts were awarded: to Great Lakes Dredge & Dock Company of Chicago (Great Lakes) for Section 1, and to Callahan Bros & Katz Company of Omaha, Nebraska (Callahan), for Section 2.

The contractors had discretion for how to dispose of excess spoil not required for filling adjoining land. Topsoil was handled separately, stockpiling it at designated locations for sale by the District, except where adjacent property owners were allowed to remove the stockpiled topsoil for spreading on their own property. In addition, bridges were built, existing sewers extended or shortened, and new outfalls constructed. In both Sections 1 and 2, the channel excavated was to be 90 feet wide with a depth 12 feet below CCD. The new channel bottom was at the same elevation throughout, with no bottom gradient. The existing bridges at Addison Street, Irving Park Road, and Montrose Avenue were moved and reused over the new channel; Great Lakes built new bridge abutments and road approaches on either side. The underside of the relocated bridge span was 17.5 feet above CCD. Temporary timber trestle bridges were provided while permanent bridges were under construction.

It is notable that the District didn't conduct soil borings and test pits to determine the character of the soil before bidding, and the contractor was required to assume all risk. Also, the contract documents didn't specify channel side slopes or require final grading and seeding to prevent erosion. The contractor was required to excavate the channel "...and shall make the sides as nearly vertical as is practical without the use of sustaining sheeting or other supports to the sides of the cut, thus allowing the sides of the channel to take a natural slope after the excavation is made." Such language wouldn't be found in earthwork specifications today; in 1904 it was a recipe for trouble.

An understanding of the mechanics of soft clays was several decades away, but the District had experience with slope stability in the earth section of the Sanitary & Ship Canal. Why that experience wasn't applied to the North Branch is inexplicable. However, dredging was eventually used throughout Sections 1 and 2, and the continual presence of water in the channel while excavation was in progress lessened the chance that slope failures would occur. Yet slope failures did occur, especially when spoil was stockpiled too close to the excavated channel. Further, the notion that the nearly vertical excavated slopes would take a natural slope when under flowing water also demonstrates a lack of understanding of river and soil mechanics, which may have been typical of the time.

The first work on both contract sections was removing vegetative cover, cutting down trees, removing stumps, and burning the woody material. Next came the horse- or mule-drawn plows and wheeled scrapers to remove topsoil, stockpiling it in designated areas. Great Lakes began their work in the vicinity of Addison Street, while Callahan began north of Montrose Avenue. Once into the clays, the excavated soil was deposited in low areas on neighboring property and used to fill in the old river bed where it was abandoned. The general elevation for filling adjacent land was 11 feet above CCD, a value determined by the District. The reach of the new channel between Belmont and Montrose was crossed numerous times by the meandering North Branch, and Great Lakes had to continually provide for flowing water moving through their work area. By contrast, the new channel north of Montrose cut across higher ground and Callahan only encountered the North Branch, and had to provide a temporary channel for it, at each end of their contract section.

As long as the surface soils were not excessively wet, Callahan found the plows and scrapers were effective at moving the dirt. However, moist soils would cause the animal hooves and the equipment wheels to bog down. Great Lakes continued with the plows and scrapers and, by July, was working throughout the Belmont to Montrose reach; it would occasionally have to shut down when rains caused the river to rise and flood the work area. Because of the meandering river and wet soils, Great Lakes used dredging once the drier surface soils were removed. Great Lakes also added timber trestles and raised the grade of the Addison, Belmont, and Irving Park bridges to increase the horizontal opening and vertical clearance under the bridges for the passage of floating equipment.

In August 1904 Great Lakes began a pilot cut with a dredge north of Belmont. The pilot cut used a dipper dredge to remove soil from the channel area and cast it to the side. Once a sufficiently sized pilot cut was completed, the dipper dredge continued, but instead of casting the spoil to the side, the spoil was loaded onto scows. The scows were towed downstream and out the Chicago River, where the spoil was deposited at Lake Front (Grant) Park or in other spoil areas under the control of the contractor. By September the pilot cut was beyond Addison, and by the time winter shut work down in mid-December, the pilot cut was almost to Irving Park. The Addison Street sewer and outfall had also been rebuilt.

Meanwhile, Callahan was encountering difficulty in Section 2. It had been assumed they would be working in the dry most or all of the time; after the topsoil was removed with plows and scrapers, they planned to use steam-powered derricks and clamshell buckets for the deeper cuts, hauling the spoil with horse-drawn carts to fill in the old river channel and low areas. In anticipation of spring floods, Callahan excavated a shallow 20-foot-wide ditch parallel to the new channel, starting from

Lawrence and progressing by September to Sunnyside Avenue. Along the route of the new channel, Callahan began to encounter soft wet clay, frustrating dry excavation by the equipment available.

By October, Callahan stopped work for the balance of the year and indicated that they were seeking another contractor to take over the work. Rather than bring in another contractor, the District agreed to assign the contract to Great Lakes and allow Great Lakes to continue by dredging all the way to Lawrence. Excavating Section 2 in the dry with equipment as the District intended and Callahan tried obviously didn't work.

Following the winter break, Great Lakes completed as much dry excavation as possible, horses and wagons hauling the spoil to fill the old river channel. In May 1905, Great Lakes completed the new Irving Park Bridge, demolished the old bridge and began moving spoil to the lake using a tug and three scows. The pilot cut was north of Irving by June and a full cut was being dredged north of Belmont. At the end of July, Great Lakes was judged to be half-done for the work south of Montrose. The pilot cut was completed to Montrose; the full cut was north of Addison by fall, and by October more bridge work was underway at Irving and Montrose. To protect against flooding south of Irving Park at the Lake View Brick Company clay pit, a levee was built in November. When work was shut down for the winter in December the contract was 70% complete.

In May 1906, an agreement between the District and the Northwestern Elevated Railroad was executed for a bridge south of Lawrence Avenue. The bridge was considered temporary and its underside had a clearance of 16 feet above CCD. In the same month Great Lakes requested additional time for removal of clay for the Bach Brick Company that was stockpiled on the west bank between Berteau and Montrose Avenues. The request was never granted and District president Robert R. McCormick, new in the job since December 1905, vetoed a Great Lakes voucher approval because, he claimed, the payment was for work performed beyond the contract completion date. The District hadn't yet dealt with the construction delays, but the trustees overrode the veto and approved the payment.

South of Montrose, Great Lakes excavated by dipper dredge due to the stiff clay. The spoil was cast into the channel behind the dipper dredge and a hydraulic dredge sucked it up and pumped it via pipeline, filling the old river channel by hydraulic sluicing. Double handling by two dredges was less expensive than loading and hauling the spoil by wagon. In May, as soon as a temporary span for the Montrose Avenue Bridge was completed and the original span removed, the dipper dredge began a pilot cut across Montrose into Section 2. Spoil was deposited in the clay pit on the east bank south of Irving Park. Dredging in Section 2 was by dipper dredge due to the depth of material and its stiffness; hence the progress was slower than in Section 1.

Great Lakes brought in two additional dipper dredges to level spoil piles near the channel and to complete the full width and depth of channel south of Irving Park. Moving these dredges required temporary disruptions in traffic for removal and replacement of each bridge. By the end of August, the new channel was complete from Belmont to Grace. The dredge in Section 2 was widening the pilot cut to full width between Sunnyside Avenue and Montrose, but north of Montrose work was shut down late in September due to the City's construction of the outfall for the Lawrence Avenue conduit. The City's contractor built a cofferdam that cut off the temporary ditch, forcing flow through the old river channel and restricting the deposition of spoil. Great Lakes focused their work south of Montrose during the shutdown, working on bridges and completing the full channel depth and width. Dredging in the vicinity of Wilson Avenue resumed in late October. Before the winter shutdown in December, dredging had progressed to north of Wilson.

In 1907 a late winter made for a slow start. The pilot cut had advanced to near Leland Avenue, but the thawing banks were sloughing into the cut. Soft material was encountered, too soft to complete the pilot cut. On the evening of February 5, both banks caved in and enclosed the dredge. Work ceased and both dredges were out for repairs. By March, the dredges were back in service dredging the full width to grade from Montrose to Sunnyside.

Full width dredging was proceeding between Sunnyside and Leland and was completed by the end of April. To avoid sloughing of the channel bank by stockpiling clay near the top of the bank, the spoil was removed by scow. North of Leland, the cut was being made by a hydraulic dredge, but only to a depth of nine feet below CCD. Between Leland and Giddings, the depth was taken by hydraulic dredge to only seven feet below CCD. The sluiced spoil was used to fill the old river channel. Hard clay was excavated by a dipper dredge north of Giddings and brought by scow to the hydraulic dredge south of Wilson, where it was pumped to fill the old channel. Some spoil was left on the east bank for the railroad's use in raising the grade of the tracks. Due to the sloughing of the channel banks it was not possible to complete the dredging by the end of July.

Great Lakes also completed bridge work at Addison, Irving, and Montrose, and all dredging was completed by August 1907. All finishing tasks were completed by November and the North Branch had a new channel. But Great Lakes had to wait almost two years, to April 1909, to receive final payment. It had been acknowledged in the September 1904 assignment of the Section 2 contract that Great Lakes would not be held to the completion time in the original contract with Callahan due to the sequential nature of the work by dredging. Although completion was 445 days late, there was no deduction for liquidated damages; the reasons given were the extra work ordered by the District and delays occasioned by the work of Great Lakes on another District contract for the Dearborn Street Bridge.

Two New Pumping Stations

The Lawrence Avenue Pumping Station operated for two decades under the ownership of the District, pumping sewage from the long narrow lakefront area after diluting it with lake water as required by the Act of 1889.

In 1921, the City was building a new relief sewer in Broadway from the south that would have required a connection to the old pumping station and additional pumping capacity. However, because of the plan for intercepting sewers and sewage treatment adopted in 1919, rather than modify the old station, the District decided to build a temporary pumping station across Lawrence Avenue at the corner of Clifton Avenue. In May 1927, the contract for the temporary Clifton Avenue Pumping Station was awarded to the T.J. Prendergast Company. The pumping station would serve only three years until after the new North Branch Pumping Station was built and in operation. Then the City's Broadway sewer could be connected to the Lawrence Avenue conduit. In 1928, Prendergast was paid an extra amount for operation of pumps and equipment to test the station before turning it over to the District, something not included in the contract.

In 1925, as part of the plan for intercepting sewers and sewage treatment for the north area, the design was underway for the North Branch Pumping Station located on the east bank of the North Shore Channel at Lawrence Avenue. In May 1926, the contract for eight electrically-driven motors and centrifugal pumps was awarded to the Allis-Chalmers Manufacturing Company.

In March 1927, a new agreement with the City updated the agreements of November 1899 and November 1910, provided for the new station and turned responsibility for the old station, including existing intake and discharge conduits, pumps, boilers, and coal handling facilities, back to the City. After the new pumping station was completed, the District constructed a connection to the existing City conduit at Francisco Avenue, installed bulkheads at the old pumping station and intake conduit, demolished the lake intake structure, and made necessary connections for City sewers to the conduit.

In December 1927, construction of the North Branch Pumping Station was awarded to Tully-Costello Company, and included all brick, concrete, and steel construction, furnishing and installing all electrical and mechanical equipment, and making connections to the Lawrence Avenue conduit and the North Side intercepting sewer. In the contract, only the pumping units were purchased separately. The contract was as close to general contracting as any awarded thus far. Tully was allowed to deposit spoil from pump station excavation on nearby District property; the result was to raise the grade of low areas making the property more useful, no doubt filling in former floodplain area. Electricity was supplied to Tully from the District transmission line and paid for at commercial rates. The new station went into service in June 1930.

Following a U.S. Supreme Court Decree, the District was planning to scale back hydroelectric generation at the Lockport powerhouse as lake diversion was reduced. In January 1929, Tully was directed to change the electrical controls and equipment to accommodate service by the Public Service Company of Northern Illinois. Later, in October, unexpected rock was encountered in constructing three 4.5-foot conduits under the channel bed connecting with the North Side intercepting sewer. In constructing the connection to the Lawrence Avenue conduit Tully had to provide temporary support for the street trolley tracks in Lawrence Avenue.

It was also part of Tully's contract to close the gates from the lake into the old pumping station intake conduit to stop the flow of water; if the gates didn't close completely, they would have to install a bulkhead. The gates did not hold against the lake, and it was determined that a bulkhead would not be successful, so in August 1930 it was decided to use clay to fill the access shaft located at the old shoreline. The intake crib was turned over to the Lincoln Park Commissioners who intended to incorporate it into park improvements. In the prior two decades the shoreline had accumulated sand, so that intake structure was at the new shoreline. In April 1927, the commissioners received a permit from the state to construct a pile bulkhead along the lake shore incorporating, and essentially eliminating, the intake structure.

The new pumping station has three dry weather centrifugal pumps, each rated at 75 cfs, which discharged sewage through a conduit under the canal to the North Side intercepting sewer. In addition, five wet weather centrifugal pumps, rated at 300 cfs each, discharged excess sewage and stormwater directly to the canal through the arched openings in the west wall of the station. The station served the same area as the former Lawrence Avenue Pumping Station; it also served additional areas in Ravenswood connected along the 2.1-mile long Lawrence Avenue conduit, a tributary area of approximately 11 square miles. In present operation, the influent sewer at the station is kept low using pumping and gravity to provide proper drainage for the tributary area. When an intense rain occurs over the tributary area, sewer water levels will rise rapidly, and by keeping the influent level low at the pumping station, excessively high water levels in the tributary area will be avoided. In dry weather it is possible to save energy by reducing pumping and allowing the influent level to rise. However, with every forecast of rain it is necessary to anticipate rising sewer levels and pump the influent level down to prepare for the increased stormwater.

North Branch Maintenance

Following completion of the new channel, occasional dredging was necessary due to two main causes: deposits of sand from Lake

Michigan brought in by dilution water; and sewage solids from the sewers and the Lawrence Avenue conduit. In March 1911, a contract was awarded to Great Lakes for dredging 188,000 cubic yards of sediment between Belmont and Lawrence Avenues. Whereas the initial excavation created a channel 90 feet wide with a bottom 12 feet below CCD, the 1911 contract called for dredging to a width of 80 feet throughout the 11,600-foot length and a depth of 13 feet below CCD from Belmont to Irving Park, and to 12 feet below CCD from Irving Park to Lawrence. Dredging was completed by the end of 1911 and the spoil deposited in designated areas of Lake Michigan at least six miles from shore.

Dredging in 1911 had a lasting effect and it wasn't until 1917 when dredging, to a limited extent, was again necessary. The commander of the U.S Army Grant Park Camp, an encampment for security during World War I, requested dredging near Roscoe Street for the storage of military vessels; the District expeditiously agreed. Great Lakes was awarded a contract to dredge a reach about 2,000 feet north of Belmont Avenue; the first 1,400 feet to Roscoe Street for a channel 20 feet wide to a depth of 15 feet below CCD and for the next 600 feet north of Roscoe for a channel 50 feet wide to the same depth. Spoil disposal was in a designated area in the lake; the work was completed in just 15 days.

Periodic dredging of the North Branch continued as more sewers were connected to the channel north of Belmont, including the North Shore Channel north of Lawrence Avenue. Not until 1928 was there any effort to remove sewage solids and the sediment in stormwater runoff before discharge to the canal; that effort began with the construction of the North Side intercepting sewer system and the opening of the North Side Sewage Treatment Works at Howard Street, renamed the O'Brien Water Reclamation Plant. All sewage from the city and suburbs north of Fullerton Avenue was redirected to the treatment works. However, during storms the North Branch Pumping Station and numerous combined sewer outfalls discharged large quantities of stormwater directly to the North Branch, and that sewage and stormwater contained solids. The frequency of dredging by the District, while less, was still continued on an as-needed basis to keep the North Branch flowing freely and to reduce offensive odors from sludge deposits.

Throughout the lengths of the channelized North Branch and the North Shore Channel downstream of the O'Brien plant, normal flow was constant as would be expected, and this steady flow was an aide in moving sediment downstream. However, three wide- or off-channel areas accumulated sediments that depleted dissolved oxygen in the overlying water: the wide areas were on the North Branch south of Diversey Parkway and south of North Avenue, and the off-channel area was the North Branch Canal. Dredging was not a sustainable practice for those areas as removal resulted in more deposition over time. The wide area south of Diversey Boulevard was created early in the 1900s after an industry moved out and a dock wall, as it appears today, was constructed. In 1938 the Julia C. Lathrop Homes project was built by the federal government landward of the then existing dock wall. In 1914 the wide area south of North Avenue was built by the Corps for a turning basin on the navigable North Branch. The North Branch Canal is discussed in Appendix A-2.

Federal Project Fizzles

Dredging became a public issue in 1945 when the Corps proposed a project to establish and maintain a 50-foot-wide, nine-foot-deep channel from Addison Street in the city to the lock in Wilmette. At Main Street in Skokie on the east bank of the North Shore Channel, the Material Service Corporation leased property from the District for a construction materials yard and dock, and had sand and gravel delivered by barge to their dock. In February 1945, a resolution of the Committee of Rivers and Harbors of the U.S. House of Representatives ignited residents and businesses along the North Branch.

The resolution was a routine request: to have the Board of Engineers for Rivers and Harbors review the reports on the proposal and determine if it was advisable to improve the channel as proposed. In May 1947 the board responded to the request, indicating that they had reviewed the reports; the board, after providing an opportunity for public input, recommended the channel be improved between Addison and Main Streets. The board also listed three locations where the channel would be widened to allow boats to pass. A local sponsor, presumably the District, would have to provide the land for the three passing places and maintain the improvements. If the governor of Illinois had no objection and Congress authorized the project, the designated federal waterway would be extended from Belmont Avenue to Main Street.

Several local organizations, including the following, protested the proposed project:

- Northwest Federation of Improvement Clubs
- Nor'Wes'Ton Congress
- Ravenswood Gardens Home Owners Association (Gardens)
- Ravenswood Manor Improvement Association (Manor)

The Gardens and Manor organizations represented home owners on the east and west sides, respectively, of the canal between Montrose and Lawrence Avenues. The protests typically claimed that the costs would exceed the benefits; taxpayers would have to pay to benefit private industry; the project was not necessary because barges and private pleasure boats already used the canal; homeowner property value would decline; current fixed-span bridges would have to be replaced with lift bridges; and sufficient commercial property existed south of Addison Street and did not need to be developed further north. It mattered not that some of these protest claims were irrelevant or not factual; the residents needed lots of ammunition.

Both organizations actively opposed the project. In October 1947, the president of the Manor organized a meeting with the Corps and representatives of elected officials at all government levels; and following, the Manor conducted a letter writing campaign opposing the proposed project. The Gardens issued a two-page detailed statement of protest in January 1948, and distributed it widely to elected officials at the federal, state, and local level. The aldermen representing the wards on either side of the river, Becker and Hoellen, became engaged and capitalized on the opportunity to grab headline notoriety. By April, Governor Green had withdrawn his support for the project and Illinois Senators Brooks and Lucas followed suit.

The Corps dropped the project, but that wasn't the last of dredging. The District issued a contract for maintenance dredging between Addison and Foster Avenues to allow freer drainage of storm water and remove a 60 to 70% blockage of the channel opposite the North Branch Pumping Station. The contract was awarded to FitzSimons & Connell Company of Chicago in October 1956 for removal of an estimated

100,000 cubic yards. In order to complete the work, the opening under the Chicago Transit Authority bridge was enlarged. The dredging was not completed until the summer of 1957 and the dredge spoil was deposited in the designated disposal area of Lake Michigan, eight miles from the shore. (Personal note: At the time I lived with my family on Giddings Street and remember watching the dredging and loading of scows. At times the odors were quite disgusting!) In September 1959, another contract for dredging was awarded to FitzSimons & Connell for removal of sediments adjacent to the pumping station deposited following a big storm on July 12, 1957.

Bridge Accident

Sludge wasn't the only material removed from the canal. In July 1957, south of Cortland Street where a branch line of the Chicago, Milwaukee & St. Paul Railroad crossed a bend in the river, a railroad bridge was pulled out of the water. The bridge had been knocked off its pier and abutment by a runaway barge that had torn away from its mooring at the Roth-Adams Fuel Company dock north of the bridge. A torrential rain caused the river to rise, slackening the mooring lines. When the lock gates at the mouth of the Chicago River were opened to release floodwater to Lake Michigan, river currents increased and pulled the barge loose. In the channel, located in the middle of a bend to the left in the downstream direction, there was a plate girder swing bridge, with the swing pier on the east bank, and a bridge seat on a pier near the west bank. With the bridge in the closed position, the barge hit the bridge broadside, knocking the west end off the bridge seat. The bridge fell into the water with a twisting motion, lifting the opposite end off the turntable; the west end plunged down into the muddy depths. Two bridge tenders on the bridge at the time were able to leap from the falling bridge onto the barge and survived the harrowing incident unscathed.

Before river traffic could pass, several days were consumed in bringing lifting equipment to the site and rigging the upended span for a lift to set it back in its open position. It took another three weeks to repair the operating machinery. The Commonwealth Edison Northwest Generating Station, located south of Addison Street, depended on frequent barge deliveries of coal to keep their generators humming and the electricity flowing, and leaving the span open was a lifeline for the company.

Placed in service in 1912, the generating station had been fitted with the largest coal-fired steam turbines of the day, two 20 megawatt units. A dependable source of cooling water was necessary, supplied by the canal and made possible by lake water diverted at Wilmette and by sewage and lake water from the Lawrence Avenue conduit. Coal was originally delivered by rail on a spur off what is presently the Metra Northwest Line, but after 1933, with the opening of the Illinois Waterway, coal was delivered by barge. A 30-day supply of coal was normally kept on hand at the generating station, but after World War II, during three actual or threatened United Mine Workers strikes, the station was converted to oil. The station was removed from service in 1970.

Stop Dumping in the Lake

In the 1960s, the environmental awareness sweeping the country arrived in the Chicago area, and the long standing practice of using the lake as a dredge spoil dumping ground became a hot issue. The District trustees, having already stopped its own practice of disposing dredge spoil in the lake, sided with leaders in the city and state government in pressuring the Corps to find locations other than the lake for dredge spoil disposal. In September 1966, the Corps capitulated and stopped using the lake; it had to find another home for the spoil by the next dredging project. Also swept up in this issue, somewhat surprisingly, was the City's practice of discharging the filter backwash water back to the lake from the two lakefront City water treatment plants. After all, the reasoning went, it came from the lake so why not put it back? However, tests showed that by concentrating what was in the lake, the sludge on the filters had high concentrations of metals and solids. Here the fix was easier; the filter backwash water was piped to the District intercepting sewers and sewage treatment plants.

In 1966, when it was time again for the District to dredge the North Branch, an unused quarry near Lemont was selected for disposal and a contract was awarded. The contract also included dredging the infamous Bubbly Creek on the South Branch. Soon after work began, the citizens of Lemont took offense that the sludge wasn't okay for the North or South Side of the city, but it was okay for Lemont. Not only was it smelly, it was alleged that it could also seep into local aquifers and pollute ground water. Over the objections of General Superintendent Vinton Bacon and Chief Engineer Ted Mickle, the District trustees canceled the contract. Since then, no more channel improvement dredging has been performed by the District due to improved treatment and solids capture.

Soft Clay and Steep Slopes

The Riverbank Neighbors, whose members live east of the North Branch between Irving Park and Montrose, describe themselves as a group of friendly, peaceful, and hard-working folks. They love their channel bank, which in their neighborhood is mostly open with access from four street ends, four alley ends, and two side alleys. Waters Elementary School is a few blocks north, and educators at the school wanted to use the neighborhood channel bank as a field site for environmental education. Appropriate permission was requested and obtained from the District to use the channel bank.

The transformation of what was a tangled mess of vegetation and a public dumping ground benefitted not only the eager learners at Waters but also the Riverbank Neighbors and the entire neighborhood. Debris was removed, undesirable vegetation was cleared out, pathways were built with logs and woodchips, and native vegetation was reintroduced. Some City services helped, but most of the grunt work was done by the Riverbank Neighbors on Saturday mornings. The District debris boat crew also assisted, dropping off cut logs for the pathways. With assistance of the Friends of the Chicago River (Friends), a grant was obtained for the installation of a fish lunker along the water line. The property line of the District was not well marked, leading to a few small encroachments, but it really didn't matter because the pathway was continuous from south of Berteau to the alley south of Montrose.

The 100-year-old channel looks like it has been there much longer. While walking along the channel several years ago, my engineer-eye noticed that the 100-year old channel seemed to be out of place. Why was the bank so steep and why was I walking so close to a nearby house? The answer to that question resulted in a District project to restore and strengthen the east channel bank between Montrose and Berteau. In the investigation leading up to the project, it was found that the slow and steady forces of moving water in the channel, working against the weaker clays in the channel bank, combined to erode the east bank; it had become much steeper and increased the risk of a slope failure, similar to what happened in 1905 and 1906.

To restore the integrity of the east bank between Berteau and Montrose, a sheet pile wall was installed, continuous except for the 100-foot-long fish lunker. The sheet pile was driven up to 30 feet into the riverbed, with the top of the wall cutoff just above the normal water line. On the channel side of the wall the sediment was minimally disturbed; on the land side of the wall, for added stability, rock backfill was placed. Soil plugs have also been placed for native vegetation to be re-established. The new, stronger channel bank is better than the rough cut of 1907 and should last another 100 years.

Before the design was finalized and permits obtained, the District met with the Riverbank Neighbors and the Friends to explain the geotechnical investigations and subsurface conditions and to review potential designs. Analysis of several soil borings suggested that simple surface treatments such as slope paving or A-jacks would not stabilize the slope. The weaker clay would continue to deform and disrupt any surface treatment. The use of a sheet pile wall was not the initial favorite in the meetings, but with the explanations of the subsoil conditions by the District geotechnical engineers, the neighbors seemed to understand the need for the structural wall.

During the time of work on the east bank, a contractor for the Corps was executing a plan for an improvement advocated by the Friends and Horner Park Advisory Council: on the west bank, flatten the side slope, remove damaged trees, refurbish the vegetation, and build footpath access to the water line of the canal. The Chicago Park District, lessee of the District channel bank, would be responsible for maintenance. Having a public park adjacent to the canal right-of-way and lacking close proximity of private residences allowed flattening the canal bank to be a viable project.

To the extent possible and practical, the North Branch, like other segments of the canal system, will be operated and maintained to meet water quality standards. Also, it will serve as an amenity for use by the public while continuing to serve its primary purpose of drainage, navigation, recreation, and stormwater management.

References

Bjorklund, Richard C., President of the Ravenswood Manor Improvement Association. Letter dated May 6, 1965, to Vincent P. Flood, Principal Assistant Attorney, District Law Department.

Bjorklund, Richard C. *Ravenswood Manor: Indian Prairie to Urban Pride*. Booklet. Chicago: Ravenswood Manor Improvement Association, 1964.

Bjorklund, Richard C. *Ravenswood Manor: Indian Prairie to Urban Pride*. Booklet. Chicago: Ravenswood Manor Improvement Association, 1981.

Brown, George P. *Drainage Channel and Waterway: A History of the Effort to Secure an Effective and Harmless Method for the Disposal of the Sewage of the City of Chicago and to Create a Navigable Channel Between Lake Michigan and the Mississippi River*. Chicago: R.R. Donnelley & Sons, 1894.

Chicago Tribune Archives. archives.chicagotribune.com.

Circuit Court of Cook County, Illinois. Chancery Division, Case No. 03 CH 21800 (consolidated with Case No. 04 CH 752 and 03 MI 718897). Consent Judgment, filed December 6, 2012.

Fink, George E. Untitled notes on the history of the Ravenswood Manor Improvement Association, c. 1950.

Friedman, John A., Acting President of the Ravenswood River Neighbors Association. Letter dated March 25, 1997, to Fred Feldman, Head Assistant Attorney, District Law Department.

Hill, C.D. "The Sewerage System of Chicago." *Journal of the Western Society of Engineers* 16, no. 7 (September 1911).

Hill, Libby. *The Chicago River: A Natural and Unnatural History*. Chicago: Lake Claremont Press, 2000.

Hogan, John. *A Spirit Capable: The Story of Commonwealth Edison*. Chicago: The Mobium Press, 1986.

Illinois Appellate Court. The Sanitary District of Chicago, Appellant, vs. The Chicago Title and Trust Company, Appellee. June 1917.

Inside Publications (Chicago). 2 September 2003.

Inside Publications (Chicago). 20 January 2004.

Larson, John W. *Those Army Engineers: A History of the Chicago District U.S. Army Corps of Engineers*. Chicago: Chicago District, 1979.

Lawrencehall.org.

Metropolitan Sanitary District of Greater Chicago. Proceedings of the Board of Trustees/Commissioners, 1955 through 1988.

Metropolitan Water Reclamation District of Greater Chicago. Contract Plans for North Branch of Chicago River Restoration Between Montrose Ave. and Berteau Ave., Contract 07-030-3D. District Engineering Department, 2013.

Metropolitan Water Reclamation District of Greater Chicago. Engineering and Maintenance & Operations Departments archives.

Metropolitan Water Reclamation District of Greater Chicago. Proceedings of the Board of Commissioners/Trustees, 1989 through 2015.

Ravenswood Gardens Home Owners Association. Meeting minutes of the Board of Managers, April 6, 1948, and June 8, 1948

Royko, Mike. *Chicago Daily News*. 21 May 1965.

Sanborn Maps. Cook County, Illinois, Jefferson Township, East half, Northwest Quarter, Section 13, Township 40 North, Range 13 East.

Sanitary District of Chicago. *Engineering Works*. August 1928.

Sanitary District of Chicago. Proceedings of the Board of Trustees, 1900 through 1954.

Wittlinger, Philip, President, Ravenswood Gardens Home Owners Association. Letter dated January 26, 1948, with attachments to the membership and residents.

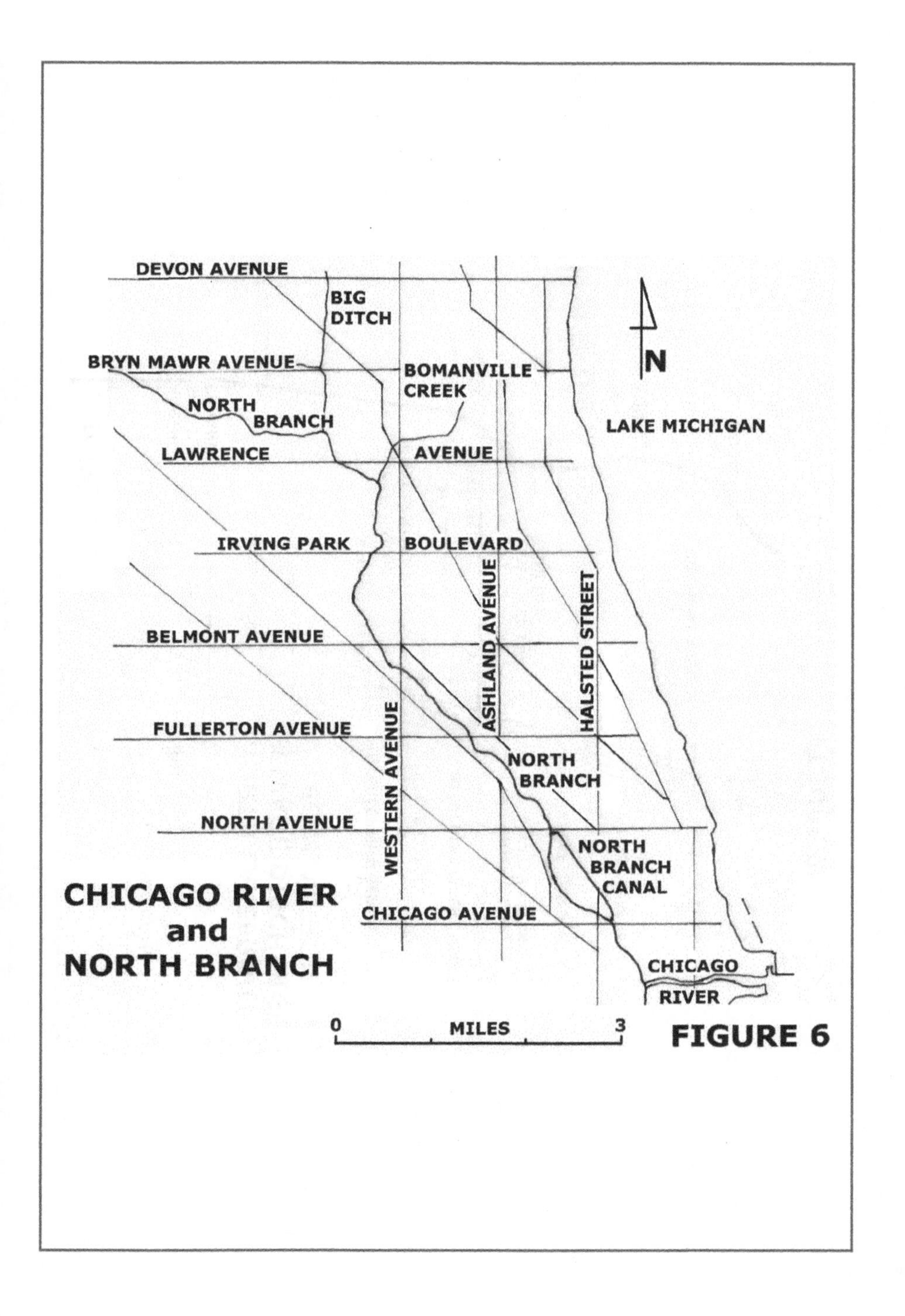
DEVON AVENUE
BIG DITCH
BRYN MAWR AVENUE
BOMANVILLE CREEK
N
NORTH BRANCH
LAKE MICHIGAN
LAWRENCE
AVENUE
IRVING PARK
BOULEVARD
ASHLAND AVENUE
HALSTED STREET
BELMONT AVENUE
FULLERTON AVENUE
WESTERN AVENUE
NORTH BRANCH
NORTH AVENUE
NORTH BRANCH CANAL
CHICAGO RIVER and NORTH BRANCH
CHICAGO AVENUE
CHICAGO RIVER
0 MILES 3

FIGURE 6

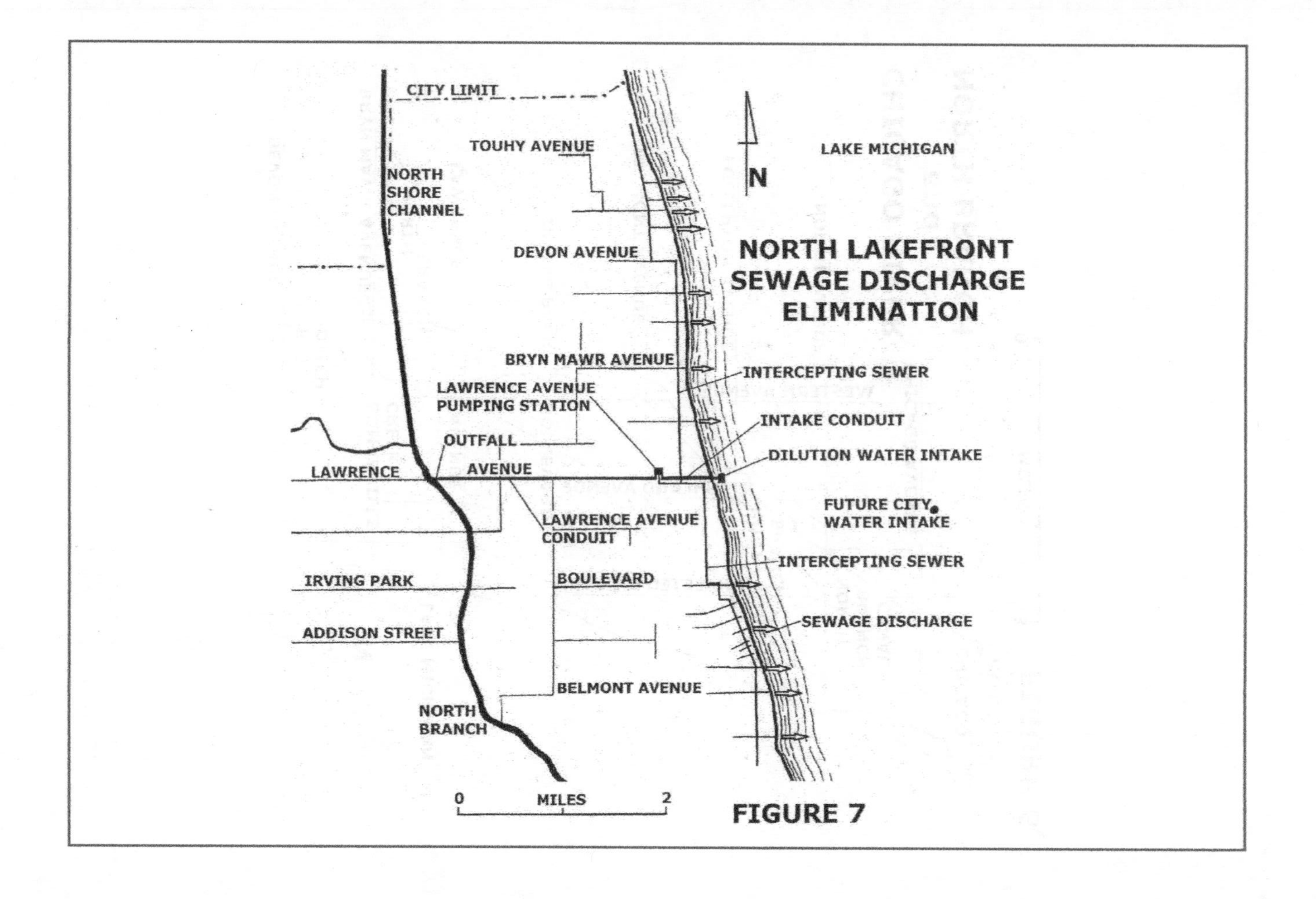
NORTH LAKEFRONT
SEWAGE DISCHARGE
ELIMINATION
LAKE MICHIGAN
N
CITY LIMIT
TOUHY AVENUE
NORTH
SHORE
CHANNEL
DEVON AVENUE
BRYN MAWR AVENUE
LAWRENCE AVENUE
PUMPING STATION
OUTFALL
LAWRENCE
AVENUE
LAWRENCE AVENUE
CONDUIT
IRVING PARK
BOULEVARD
ADDISON STREET
BELMONT AVENUE
NORTH
BRANCH
0
MILES
2
INTERCEPTING SEWER
INTAKE CONDUIT
DILUTION WATER INTAKE
FUTURE CITY
WATER INTAKE
INTERCEPTING SEWER
SEWAGE DISCHARGE

FIGURE 7

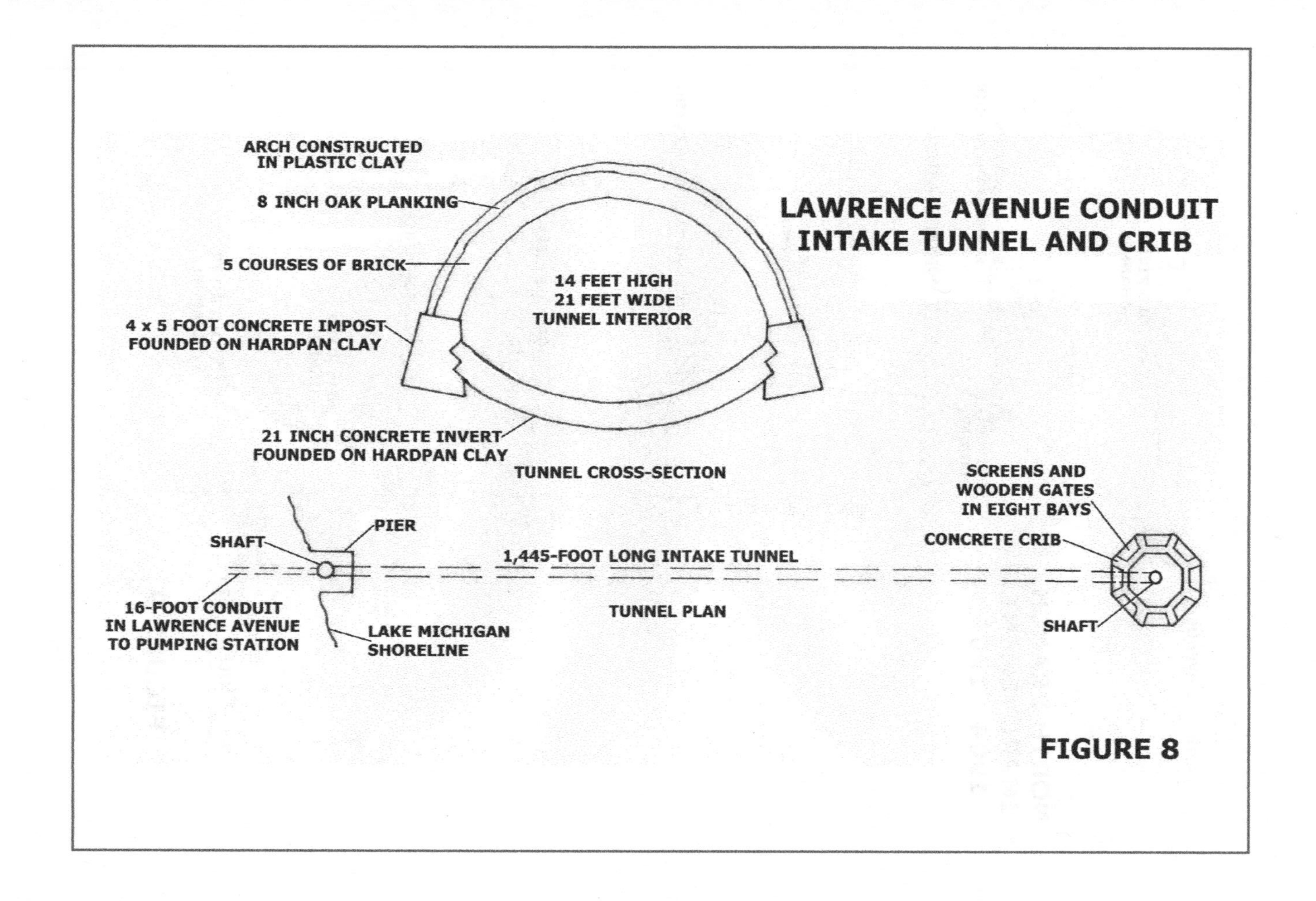
LAWRENCE AVENUE CONDUIT
INTAKE TUNNEL AND CRIB
ARCH CONSTRUCTED
IN PLASTIC CLAY
8 INCH OAK PLANKING
5 COURSES OF BRICK
4 x 5 FOOT CONCRETE IMPOST
FOUNDED ON HARDPAN CLAY
14 FEET HIGH
21 FEET WIDE
TUNNEL INTERIOR
21 INCH CONCRETE INVERT
FOUNDED ON HARDPAN CLAY
TUNNEL CROSS-SECTION
SHAFT
16-FOOT CONDUIT
IN LAWRENCE AVENUE
TO PUMPING STATION
PIER
LAKE MICHIGAN
SHORELINE
1,445-FOOT LONG INTAKE TUNNEL
TUNNEL PLAN
SCREENS AND
WOODEN GATES
IN EIGHT BAYS
CONCRETE CRIB
SHAFT

FIGURE 8

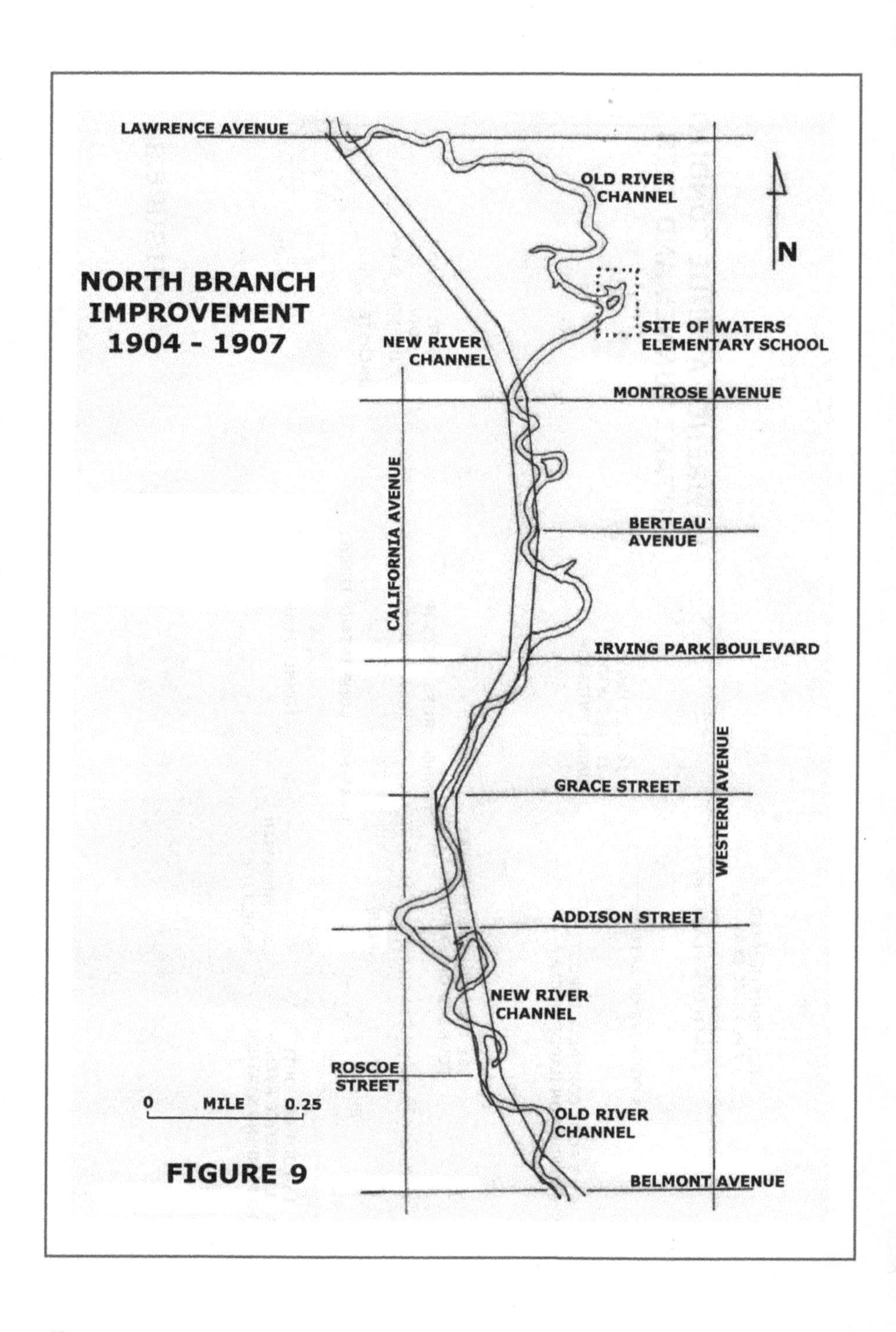
LAWRENCE AVENUE
OLD RIVER CHANNEL
N
NORTH BRANCH IMPROVEMENT 1904 - 1907
SITE OF WATERS ELEMENTARY SCHOOL
NEW RIVER CHANNEL
MONTROSE AVENUE
CALIFORNIA AVENUE
BERTEAU AVENUE
IRVING PARK BOULEVARD
WESTERN AVENUE
GRACE STREET
ADDISON STREET
NEW RIVER CHANNEL
ROSCOE STREET
0 MILE 0.25
OLD RIVER CHANNEL
BELMONT AVENUE
FIGURE 9

Photograph 2.1: June 29, 1922. The Lawrence Avenue Pumping Station was completed by the Chicago Department of Public Works in 1906 and turned over to the District in 1910 for operation and maintenance. Located on the north side of Lawrence Avenue immediately west of the present day Chicago Transit Authority Red Line, it pumped sewage and lake water west to the North Branch through the Lawrence Avenue conduit. Today, the site is a municipal parking lot. (MWRD photo 9234)

Photograph 2.2: Looking southwest along the North Branch toward Montrose Avenue circa 1900, prior to its relocation by the District. The river channel turns to the east in the distance as it passes under the Montrose Avenue Bridge. Large, flat, forested floodplain areas paralleled the stream channel. Except for buildings facing Lawrence Avenue, development north of Montrose Avenue had not yet occurred. (Chicago Public Library photo RLVCC 1.1069)

Photograph 2.3: This expansive north-looking view of the North Branch near and beyond Lawrence Avenue on April 8, 1903, illustrates the low stream banks and broad flat floodplain areas on each side of the natural river channel. Each end of the recently built bridge stops abruptly as the road approaches have not been built up with fill and graded. Built by the City, the bridge was completed by the District and opened to traffic in late 1909. (MWRD photo 2315)

Photograph 2.4: The first step in construction of the North Branch improvement was the removal of topsoil, here being performed on May 21, 1904, with a motor-driven grading machine, which scrapes off the topsoil and loads it into a horse-drawn wagon. The wagon load was transported to a designated location for stockpiling topsoil for later use or sale. Much of the topsoil was given to local land owners as a condition of the purchase of their land by the District. (MWRD photo 2753)

Photograph 2.5: Belmont Avenue crossed the North Branch with a through truss center pier swing bridge, shown here on October 14, 1903. The channel improvement by the District started just north of this bridge, which was also the upstream limit of channel maintenance by the Corps of Engineers. This was the last movable bridge on the North Branch and provided boat passage for a boatyard on the west bank north of the bridge. The trees behind the bridge are the location of the picnic grove in Riverview Amusement Park due to open in 1904. (MWRD photo 2483)

Photograph 2.6: April 8, 1903. A west-bound, horse-drawn wagon is crossing the Irving Park Road Bridge. The masonry bridge abutments and the bowstring truss span crossing the channel were replaced as part of the improvement of the North Branch. Sewer outfalls in or alongside the bridge abutments were also rebuilt to accommodate the new wider channel. (MWRD photo 2321)

Photograph 2.7: May 21, 1906. A temporary Irving Park Road Bridge is in place over the new widened channel. The original masonry bridge abutments have been removed and replaced with timber pilings and the original bowstring truss span has been replaced with a Brown-type truss span on the timber abutments. The temporary truss span can be easily removed and replaced for passage of large floating equipment used for dredging the new channel. (MWRD photo 3409)

Photograph 2.8: The east masonry abutment of the Montrose Avenue Bridge includes a large diameter sewer outfall underneath the bowstring truss span. Large houses are in the neighborhood south of Montrose and east of the channel on April 8, 1903. The North Branch is flowing toward this bridge from the northeast in the left foreground, but the new channel will be cut to the northwest and will flow toward the bridge from the right foreground. (MWRD photo 2317)

Photograph 2.9: May 21, 1906. Looking east over Montrose Avenue from a nearby rooftop, two bridges are visible: the original bowstring truss at left and the abutments for the temporary bridge on the far right. Planking for the road diversion crosses the foreground. (MWRD photo 3411)

Photograph 2.10: Looking north from the Belmont Avenue Bridge over the North Branch on April 7, 1905, the initial cut made by a dipper dredge appears to the right, with the original channel on the left next to the boatyard dock wall. The spoil between the two channels will be removed later when the final dredging is completed. The large building and smokestack in the background is believed to be an early electrical generating station. (MWRD photo 3079)

Photograph 2.11: A dipper dredge is at work south of Roscoe Street on April 7, 1905, widening the initial cut to allow the passage of scows and tow boats. The Belmont Avenue Bridge is in the left background and the housing in the right background is a development called Electric Park. (MWRD photo 3080)

Photograph 2.12: A hydraulic dredge is working north of Addison Street on May 21, 1906. The cutter head is raised just above the water level indicating that the dredge is adjusting its position in the channel before lowering the cutter head and continuing to excavate. The cutter head arm rotates on a vertical axis allowing the cutter head to sweep from side-to-side when excavating. The spoil slurry is discharged through the pipe behind the dredge leading to a disposal area on the west side of the channel. (MWRD photo 3418)

Photograph 2.13: This area east of the North Branch will receive hydraulic dredge spoil by sluicing sometime after April 17, 1907. Ground vegetation is still evident and laborers at right are placing fill to build a containment berm for the sluiced sediment and water. Standing water in the spring on this floodplain area is common resulting from North Branch overflow and/or snowmelt. (MWRD photo 3676)

Photograph 2.14: Looking southwest on June 27, 1907, from a point south of Wilson Avenue along Campbell Avenue, an area already subdivided with dedicated streets and alleys. The containment berm on the left envelopes the site of what will be Waters Elementary School in four years. Some of the trees in this view may be those still standing on the south end of the school campus. The author attended this school in the late 1940s and remembers the custodial staff complaining of seepage in the lower level boiler room. (MWRD photo 3741)

Photograph 2.15: Looking southwest over the disposal area from the corner of Leland and Maplewood Avenues on June 26, 1907. The containment berm is to the left and ends at a concrete wall, then continues beyond the wall. The wall is part of the structure where the descending elevated railroad approaches grade. The tracks are seen sloping downward from left to right and are supported on timber pile bents above the sediment. After the sediments consolidate and dry out, crushed rock fill will be placed to support the tracks of the present Chicago Transit Authority Brown Line. (MWRD photo 3744)

Photograph 2.16: Looking northwest from the corner of Leland and Maplewood Avenues on June 26, 1907. The containment berm is seen to the right. Recent sluicing is indicated by the edge of deposition to the left of the berm. Buildings along Lawrence Avenue are in the background. To the right of center is a sign next to the berm. It reads "WARNING: This dam is for temporary use during filing progress by hydraulic dredge. Persons interrupting this dam will be prosecuted according to law. SANITARY DISTRICT OF CHICAGO." (MWRD photo 3745)

Photograph 2.17: The end of the pipeline discharging the dredged spoil slurry into the disposal area is on the right in this June 26, 1907, view of the disposal area looking east south of Lawrence Avenue. In the foreground is the containment berm to restrict the spread of sediment. The clay will settle and the water is returned to the new channel. (MWRD photo 3748)

Photograph 2.18: The outfall of the Lawrence Avenue conduit on the east bank is discharging sewage diluted with lake water on July 12, 1909, from the Lawrence Avenue Pumping Station two miles to the east. This view from the west bank shows part of the unfinished Lawrence Avenue Bridge and the west end of the Lawrence Avenue trolley line. Sewage no longer discharged to the lake along the shoreline from the Chicago River north to Calvary Cemetery. (MWRD photo 4111)

Photograph 2.19: Looking south from the east bank at approximately Leland Avenue on January 15, 1907, bridge construction is in progress for the crossing of the Northwestern Elevated Railroad, later called the Chicago Transit Authority Ravenswood Line and presently the Brown Line. The dipper dredge is excavating near timber pilings opposite the abutment on the west bank. (MWRD photo 3575)

Photograph 2.20: A Ravenswood Manor neighbor has driven up to inspect the repaired barricade on Sunnyside Avenue on June 8, 1917, where a taxi cab crashed through the barricade and into the channel. Across the North Branch in the Ravenswood Gardens neighborhood, houses along Virginia Avenue backup to the channel. Both neighborhoods developed rapidly prior to World War I. The Wilson Avenue Bridge is visible behind the street light. (MWRD photo 5975)

Photograph 2.21: This northward look from the Belmont Avenue Bridge on May 27, 1920, shows the North Branch busy with the Riverview Amusement Park on the right bank and the Grebe Boat Yard on the left bank. The Commonwealth Edison Northwest Electrical Generating Station with six smokestacks is in the background. Coal was initially brought to the station by rail, but with the opening of the Illinois Waterway in 1933, coal was delivered by barges coming up the North Branch. (MWRD photo 7541)

Photograph 2.22: Across the channel north of Lawrence Avenue is the site of the soon to be constructed North Branch Pumping Station on February 26, 1925. Lawrence Hall School for Boys built a dozen years earlier stands east of the site facing Francisco Avenue. (MWRD photo 11561)

Photograph 2.23: The contractor building the North Branch Pumping Station was allowed to deposit spoil from the excavation of the foundation on District land across the channel and north of Lawrence Avenue, shown here on March 29, 1928. The filled-in floodplain, presently Ronan Park, is leased to and managed by the Chicago Park District. (MWRD photo 13843)

Photograph 2.24: The depth of the Lawrence Avenue conduit required a deep foundation for pumps at the North Branch Pumping Station. Four pairs of suction bells for the large stormwater pumps are shown to the right on June 19, 1928, being cast into the concrete foundation. To the left are suction bells for the smaller dry weather sewage pumps. The influent sewer screen chamber is at the far south end of the structure. (MWRD photo 14065)

Photograph 2.25: Viewed from above in Lawrence Hall across the street, the pumping station substructure is nearly complete on October 26, 1928. The screen chamber is at left, five holes in the channel wall are for the stormwater pump discharge pipes, and a derrick has been set up to begin erection of the structural steel superstructure. (MWRD photo 14429)

Photograph 2.26: Unexpectedly, rock was encountered when tunneling under the channel for the three-barrel inverted siphon connection to the North Side Intercepting Sewer northwest of the pumping station. The laborers are taking a break on October 26, 1929, while excavating through the rock. (MWRD photo 15956)

Photograph 2.27: Nearing completion of pump installation on March 29, 1930, four of the five wet weather pumps and two of the three dry weather pumps and their electrical motors have been installed. The design of the station included flexibility in managing pumped flow through multiple connections and gate chambers. In addition to pumping from the Lawrence Avenue conduit, the station can also pump from the intercepting sewer west of the channel, thus expanding the area tributary to the station for relief of excess stormwater. (MWRD photo 16237)

Photograph 2.28: Four months before placing the pumping station in service, most of the structure of the pumping station is complete on February 25, 1930. Yet to be completed is the masonry work on the south end covering the screen chamber, landscaping, and various controls and instrumentation of the electrical and mechanical equipment. (MWRD photo 16137)

Photograph 2.29: April 15, 2012. After 82 years of service, the North Branch Pumping Station is as good as new, a tribute to careful maintenance and periodic replacement of worn pump impellers. Artful use of color lends an air of liveliness and brightens the outlook of employees and visiting public, a contrast to past drab monotone color schemes. (Photo courtesy of Lindsay Olson)

Photograph 2.30: A view from the Montrose Avenue Bridge on April 26, 2015. The new east wall and embankment of the North Branch Channel stretches one-quarter mile to Berteau Avenue, replacing what was an eroded and dangerously steep earthen slope. After more than 100 years of continuous operation, the North Branch, like the other District canals, is good for another century of service. (Photo by the author)

Chapter 3

North Shore Channel

Introduction

The Act of 1903 annexed the area north of the City of Chicago to the county border and provided for an outlet for sewage from the growing North Shore suburbs. Evanston and other communities had opted out of the District when the boundaries were established in 1889, but in just a few years attitudes had changed. Glencoe constructed a large septic tank to address its need for sanitation. However, this treatment technology was limited to small populations. Even with an effective septic system, there was no other outlet for the effluent than Lake Michigan.

The suburban area was served with a few drainage ditches implemented by municipalities or with drainage districts under the agricultural drainage district statute. The population had become large, overwhelming the natural systems, so the ditches were no longer suitable as outlets to receive septic tank effluent or raw sewage. Incorporating these growing suburbs into the regional District was the sensible thing to do. For some towns and Lake Michigan, relief couldn't come soon enough. However, before construction began, the channel was controversial. An early plan by Evanston had the channel protruding well into Wilmette, and Wilmette objected strenuously because the proposed channel would displace some of its more affluent residents. Proponents in the city suggested that the channel be funded by special assessment of the northern suburbs; the northern suburbs opposed this approach, instead favoring that the cost be borne by the entire District tax base. In the Act of 1903, the latter prevailed; special assessment was not included in that act. Suburban opponents proposed that the expansion of the District territory be decided by referendum, believing it would be defeated. However, the General Assembly only required a referendum if enough voters signed petitions. A petition drive didn't occur.

Surveys and design of the new channel began before the statute was amended and by November 1903, the public was informed of the schedule of land acquisition and construction, and the benefits to be achieved by the new channel. This publicity was proactive to counter opposing views favoring a referendum on annexation.

What was initially known as the Evanston Channel was eventually constructed and connected to the North Branch near Lawrence Avenue in Chicago. The improvement of the North Branch between Belmont and Lawrence Avenues provided an ample outlet for the new channel to convey the sewage from the north suburbs and dilution water pumped into the new channel from Lake Michigan at Wilmette.

Where the North Branch flowed into the new channel south of Foster Avenue, a low dam was needed to protect and preserve the natural channel of the North Branch upstream of the confluence. The improvement of the North Branch downstream of Lawrence Avenue lowered the water level a few feet and without a dam to preserve the natural channel, considerable erosion and down-cutting of the natural channel would occur and the District would have been liable for damages to structures and loss of property value.

Construction of the North Shore Channel brought its own set of challenges, despite the experience gained by the District on its prior projects. The sloughing of channel banks experienced on the dry excavation of the earth section of the Sanitary & Ship Canal and the wet dredging of the North Branch improvement were minor compared to the problems yet to come.

Route Selection

Route selection was a no-brainer. As explained by Libby Hill in *The*

Chicago River: A Natural and Unnatural History, the route followed an older drainage ditch. The older ditch had become neglected and filled with sediment; it needed to be made larger to carry away sewage and dilution water. Route selection was consummated in July 1904, slightly different than the route recommended by the special commission of experts in 1901. The route crossed three railroad lines and the District retained the services of John Harlan as special counsel for railroad negotiations, a departure from the past practice where the law department attorneys handled these negotiations. (See Figures 10 and 11.)

The Village of Winnetka requested a hearing on the route of a channel to express their desire for a channel near to the village that could be used as an outlet for drainage at less cost and without the complication of other intervening municipalities. Winnetka was obviously concerned that neighboring municipalities may be difficult and obstructionist when it came time to approve requests to pass though the neighboring jurisdictions. A formal hearing was never held.

Land acquisition and planning began in 1904 requiring more field surveys for plats and legal descriptions. Many properties contained commercial or residential buildings and the engineering department assisted with structural integrity assessments and cost estimates for relocation. The July 1904 ordinance was modified twice in 1905 to make minor adjustments in route details and it was hoped that the revision would allow one bridge to be built in the city for Lincoln and Peterson Avenues rather than two. However, the City wouldn't agree. One more attempt was made four years later to build only one bridge for these two thoroughfares and again the City refused.

The election in November 1905 dramatically changed the composition of the board of trustees and at the last meeting of the old board the clerk was ordered to gather all appraisals for the property to be acquired so that this information would be immediately available to the new board, which took office the following day in December. Although a helpful gesture, it proved fruitless as the new board under the leadership of President Robert R. McCormick soon demonstrated that it wanted little to do with the work of the prior board, a change from past practice.

Land Acquisition

Much of 1905 was occupied with design and surveys to prepare for the new channel: property line surveys, soil borings, meetings with railroads for three bridges, meeting with municipalities and township road commissioners for 22 road bridges, working with municipalities and utility companies on utility relocations, design of temporary rail and road diversions and bridges, preparing plans and specifications for the channel excavation contracts, and design of lateral drainage ditches. Very little land was acquired.

In January 1906, the new president reminded the trustees of promises made during the campaign to move expeditiously on construction of the Evanston Channel, but then suggested that past decisions be reviewed. He presented an ordinance for consideration that called for the repeal of some parts and retention of other parts of the earlier ordinance defining the channel route. The trustees were in favor of repeal of the entire ordinance, but urged early consideration of a modified route and construction of the channel. Not until May was any concrete action taken, and that was a direction to the chief engineer to report at the next meeting on the plans for the channel and the amount of land to be acquired. The report was never given, perhaps because of subsequent events. Meanwhile, an automobile was hired for the trustees to inspect the proposed route of the channel, but the inspection trip, if it occurred, was never acknowledged or reported publicly. In spite of the campaign promises and the readiness of the engineering department, the trustees took their time in making decisions.

The trustees wanted to proceed with land acquisition expeditiously within the established right-of-way while they were considering the size and other details of the constructed channel. However, there were differences of opinion among the trustees regarding the amount of land to be acquired. Some wished to acquire the full width in the established right-of-way, others wished to acquire the full width only at the north end between Lake Michigan and the Chicago & North Western Railroad in Evanston. A third faction wished to acquire the full right-of-way plus an additional strip 385 feet in width to the east of the established right-of-way between the Chicago & North Western Railroad and Lawrence Avenue. This strip was for the purpose of stockpiling the excavated spoil. In an apparent move to be helpful and provide direction, more than one resolution was presented for consideration, but the existing ordinance was not replaced.

Opinions were desired on the title of property within the proposed right-of-way, but it wasn't until September that the trustees decided to enter into an agreement with Chicago Title & Trust Company. Also, the chief engineer was directed to prepare maps and legal descriptions of all parcels within the right-of-way for the channel for use by the title company and in condemnation proceedings. The following month, the trustees moved to provide funds for property acquisition. Numerous parcels of land were acquired from 58 individuals in 1907. A local surveyor in Wilmette was hired in February to locate block corners in the village within the channel right-of-way because it was believed that a local surveyor was more familiar with local land rights and surveys and could do the work at less cost than District surveyors.

In March, Evanston adopted an ordinance granting certain rights to the District and vacating alleys and streets within the established right-of-way for the channel; the District readily accepted the Evanston ordinance. The Evanston ordinance acknowledged that the construction of the channel will provide an outlet for their sewage, thereby preserving the purity of their water supply. The ordinance required that the spoil from excavation of the channel east of the Chicago & North Western Railroad be deposited outside the city limits, named all streets and alleys to be vacated, named bridges to be built and their specifications, and contained other details regarding traffic control, utility relocations, and sewer and water main replacement. It is noted that Evanston didn't restrict the deposition of spoil west of the railroad even though the channel right-of-way was within the city limit as far as Emerson Street and continued to abut the city limit as far as Crain Street.

John Emerson, an owner of property along the route of the channel, obtained permission from the District in July to install a tile drain from his property across District land to the point where the channel was to be built to allow drainage from Emerson's property to continue after the placement of spoil, which would eventually block the swale. Emerson owned land west of the channel route in the vicinity of what was to become the intersection of McCormick Boulevard and Golf Road, an extension of Emerson Street.

Appraisers were retained in July to provide new appraisals as the prior appraisals had aged and were of little value in property acquisition. The firm of Coats & Burchard was hired to appraise buildings and three individuals, William A. Bond, John C. Fetzer, and Edwin F. Getchell, were hired to appraise land. Later in October, the District's real estate manager began to sell the topsoil and trees on the channel right-of-way, including the property purchased for spoil disposal. Another right-of-way change in August adjusted the route between Lawrence and Foster Avenues in the city so that the owners of property along the east side of Kedzie Avenue would have more depth to their holdings, increasing the lot depth from 124 to a more normal 186 feet.

The National Brick Company (National Brick), situated along the channel route near Touhy Avenue in the city, was interested in using clay from the excavated channel for making bricks. Agreement was reached in August 1907 for the area from Pratt Avenue to Emerson Street and was basically a swap of land for spoil. The agreement provided that National Brick would:

- convey a strip of land 395 feet wide by 0.5 miles long containing 24 acres for the new channel,
- excavate a portion of the channel 80 feet wide, 16 feet deep, and 1.0 miles in length by January 1, 1910, and
- remove 100,000 cubic yards of spoil each year, completing the removal of spoil from the spoil piles by January 1, 1929.

The District would:

- sell all spoil from the channel construction to National Brick from Pratt Boulevard to the Chicago & North Western Railroad Milwaukee Division, and
- lease to National Brick the easterly five feet of the property conveyed for 30 years without rental with an option to extend for an additional 20 years at rental based on appraised value.

In April 1910, a supplemental agreement with National Brick was approved providing for a five month time extension to complete the removal of clay from the spoil piles and allowing National Brick to construct a dock on the channel, the first indication of commercial navigation use for the channel.

The channel route crossed the right-of-way of the Chicago & North Western Railroad at two locations, near West Railroad Avenue, later renamed Green Bay Road, and Oakton Street. Even though the engineering departments of the railroad and District were discussing an agreement for the bridges and channel to be constructed, the trustees directed the attorney to initiate condemnation. An agreement was eventually approved that allowed the railroad to build both bridges, the District to reimburse the railroad and required the track elevations to comply with the Evanston ordinance.

Negotiations were also in progress on the many road bridges. In October, agreement was reached with the Highway Commissioners of Niles Township for the crossing of the channel for six roads—Oakton Street and Touhy Avenue, and Church, Dempster, Emerson, and Lincoln Streets. The agreement allowed the District to install towers and wires to transmit electricity, cross the roads with railroad tracks, and operate a railroad, if such were necessary for channel construction. The agreement required the District to maintain the permanent bridges in perpetuity and to provide temporary bridges while permanent bridges were being built.

A lease was executed in November 1907 with Theodore and Henrietta Schramm for the two-story building at 1600 Noyes Street, Evanston, until April 1908, conforming to the purchase agreement for the property. The Schramms continued to operate their business on the ground floor and rent the upstairs apartment for the term of the lease at which time it was expected that channel construction would begin. Similar leases were executed for short-term occupancy to other parties. Month-to-month leases were signed the following spring for five residences in Evanston on Dewey and Noyes Streets and the real estate manager was directed to sell houses located on the newly acquired land at public auction. Fifteen houses on Cooper, a vacated street, and Darrow, Dewey, Noyes, and Wesley Streets were listed for sale as well as one on the former Evanston Golf Club property. A greenhouse, including all buildings and machinery, located on Grey Avenue between Foster and Simpson Streets was also sold.

Construction of the channel near the lakefront had already begun late in 1907 and the pace of land acquisition quickened. In 1908, 86 separate land purchases were processed. The right-of-way ordinance was again revised in February to correct certain property names and descriptions, change the route south of Touhy Avenue in Chicago, add land between Lincoln Street in Evanston and Hill Street, later renamed Maple Avenue, in Wilmette, and add lots in Wilmette near Lake Michigan. The latter change was made as design of the Wilmette Pumping Station and Sheridan Road Bridge progressed. Special assessments were paid to the City of Evanston, for sewers in Grant Street and the paving of Darrow Avenue. These improvements were public works projects begun before the property was acquired by the District.

A Village of Wilmette ordinance adopted in March 1908 vacated certain streets (Central, Greenleaf, Laurel, and Michigan Avenues; Sheridan Road and Second Street) and alleys within the channel right-of-way; granted the right to construct the channel through the village; retained some streets and alleys for the eventual construction of relief sewers by the village; identified the streets that must be dedicated and paved by the District; required the District to build temporary and permanent bridges for Hill, Laurel, and Linden Streets and Sheridan Road; prohibited the District from permanently stockpiling excavated spoil within the village; required the pumping station to be constructed north of Greenleaf Avenue and to be powered electrically from a source outside the village; required the District to provide for the support of all water mains and other utility services crossing the channel right-of-way, including the water mains that supply water to the village from Evanston; and required that all sewers and drains encountered in the construction of the channel must be properly re-routed.

The Evanston Golf Club leased land to the District for a railroad siding east of the Chicago, Milwaukee & St. Paul Railroad, presently the Chicago Transit Authority Purple Line, and south of Hill Street. The term of the lease was for 15 months, but was later extended for two years while construction was still in progress.

Some land acquisitions were not so simple. The District eventually decided to acquire the entire lot owned by Eliza A. Wheeler rather than just the fragment of the lot within the established right-of-way for the channel. Upon the initial offer for the fragment, Ms. Wheeler refused to negotiate the sale of the fragment and would only negotiate if the entire lot were purchased. The three appraisers were consulted and agreed that it was advisable to purchase the entire lot because the purchase price of the fragment plus the damages to the remainder would cost more than purchase of the entire lot. *Damage to the remainder* is a term of art in the taking of property; it is the economic value of losses the property owner faces for the diminished size of their holding. Examples are loss of income from the property taken and increased cost for access to the property not taken. The opinion of the appraisers held and Ms. Wheeler got her way. This property is in the vicinity of the present day Evanston Community Gardens at McCormick Boulevard and Bridge Street. These transactions kept the engineering department busy updating plats of survey and the right-of-way map and also resulted in the District owning bits and pieces of property not needed for construction.

William A. Peterson owned considerable property along the route, and as a condition of the purchase of some of his property, he wanted the spoil removed from the sold property to be deposited on his other property south of Bryn Mawr Avenue in Chicago, where he operated a flora nursery. Peterson wanted to fill in low areas and increase the value of his land for development. The purchase agreement also committed the District to dedicate and improve Kedzie Avenue between Bryn Mawr and Lincoln Avenues. Meanwhile, the District offered to donate land to the City for the re-routing of Lincoln Avenue via Peterson Avenue and a new road along the west side of the channel connecting Peterson and Lincoln Avenues, thereby saving the cost of a bridge for Lincoln Avenue. An ordinance to accept the offer and permit the road relocation was pending in the city council, but it was never enacted.

The District also realized that dedicating and improving Kedzie Avenue would interfere with the channel route, so dealing with Peterson came back into the picture again in October 1909. In another agreement, the District dedicated and improved a 66-foot-wide strip of land on the west edge of the channel right-of-way from Bryn Mawr Avenue to Lincoln Avenue for a new public right-of-way and street. The new street replaced Kedzie Avenue and the improvement was for a 16-foot wide macadam pavement in the center of the 66-foot dedication. The agreement also committed Peterson to join with the District in seeking to have the City vacate Kedzie Avenue between Bryn Mawr and Lincoln Avenues and to accept the dedication of what was then called Spaulding Avenue. If the City agreed, then Peterson would release the District from its obligation in the August 1908 agreement for the dedication and improvement of Kedzie Avenue between Bryn

Mawr and Lincoln Avenues. Peterson also agreed to remove the trees and topsoil from the new right-of-way. However, the District was not successful in having the City agree to eliminate the Lincoln Avenue Bridge. The construction of the road on north Spaulding Street from Bryn Mawr to Peterson in the city was completed in September 1910. What was then referred to as Spaulding Avenue is presently Jersey Avenue. (See Figure 12.)

The District needed an office and sleeping quarters for the police department during the construction of the channel and entered into a lease with the Chicago, Milwaukee & St. Paul Railroad in November for the unused and vacant railroad station at Llewellyn Park. The building was located at Third and Hill Streets in Wilmette. The building was vacant because the railroad had been extended earlier and a new station built at Linden Street, two blocks north.

The Illinois Brick Company owned a clay pit located northeast of the corner of Lincoln and Peterson Avenues, which they desired to have filled with spoil from the excavated channel. A contract in February 1909 between the brick company and the District accomplished the filling and allowed the District to install tracks, switches, turnouts, and sidings as may be necessary to haul spoil to the clay pit. Filling the pit was a long-term matter. In October 1915, a supplemental agreement allowed the Great Lakes Dredge & Dock Company (Great Lakes), a District contractor, until May 1916 to complete the filling with spoil from dredging the channel and from other sources. The obligations of the prior agreement were released by mutual consent.

In 1909, 46 individual parcels of property were acquired for channel construction, among which was a parcel on the south side of Grant Street in Evanston, east of the channel centerline in Evanston, owned by the McCormick Estate. Mr. R. Hall McCormick was the trustee of the estate and President McCormick declared for the record in April 1909 that he was abstaining from voting because the trustee was his second cousin.

Clara F. Bass of Peterboro, New Hampshire, owned a strip of land along one mile of the channel north of Foster Avenue in the city and to avoid prolonged condemnation litigation in May 1909, Joseph J. Budlong and James H. Van Vlissingen were hired to act as mediators, to appraise the value of the properties to be purchased and the damages to the remainder. Both men were familiar with the value of properties in the area, Budlong owning much land used for truck gardening and Van Vlissingen acting as an appraiser and arbitrator for land sales throughout the Chicago area. They recommended that a three-party agreement be entered into between Bass, the District, and William A. Peterson, to provide for the dedication of a 66-foot-wide right-of-way for Bryn Mawr Avenue from the east line of Kedzie Avenue to the east line of the Bass and Peterson properties, the construction of a bridge by the District over the channel at Bryn Mawr, the lease to Bass of a 300-foot frontage along the east side of the channel extending to Foster Avenue with a specified annual rental and term, and the District to connect the Chicago & North Western Railroad spur track serving the property of the J.J. and L.A. Budlong Company.

The connection of the railroad spur, the lease to Bass, and the building of the bridge all served to significantly reduce the damage to the remainder. Further, the building of the bridge and lease to Bass was a condition for her agreement to sell property to the District. Among the conditions of the Bass lease, two were of interest: allowing Bass access to the channel for commercial use, such use being not in conflict with the use of the channel by the District, and allowing Bass to use the lease premises for roads, street railways, or railroad routes, where these uses are prohibited within 40 feet of the channel. Among the conditions of the agreement was a provision that the bridge be sufficient to carry a street railway. With all these inducements, Bass sold and the District avoided condemnation. The spur track mentioned was an extension of the Chicago & West Ridge Railroad. Part of the Bass holdings was located adjacent to the Chicago & North Western Railroad south of Oakton Street. (See Appendix A-5.)

In August 1909, a land swap and exchange of deeds between the District and three parties improved the alignment of Sheridan Road for the location of the bridge and pumping station. Corinne True, Arthur S. Agnew and his wife, and Bernard M. Jacobson and his wife deeded to the District 3,960 square feet of property. In return, the District deeded to them 4,572 square feet for the Baha'i Temple Unity Corporation. Both pieces of property were appraised at roughly the same value and no money was exchanged. In October, the District portion was dedicated to the Village of Wilmette for the changed location of Sheridan Road and Michigan Avenue. The Village of

Wilmette accepted the swap in November 1910 memorializing the relocation of Sheridan Road and Michigan Avenue. (See Figure 13.)

A claim for damages resulting from the vacating of streets in Evanston by rededicating the vacated streets at Colfax Street and Jackson Avenue was settled with a homeowner in November. In February 1911, the District assumed the defense of two claims by property owners on Dodge Avenue and paid damages to settle the claims. The same occurred in March for a property on Darrow Avenue. Both were handled by the District because the March 1907 ordinance of the City of Evanston required the District to hold the City harmless for any claims by property owners. In June, the District dedicated a strip of land for the creation of Bridge Street, connecting the intersection of Payne Street and Grey Avenue with the intersection of Brown Avenue and Simpson Street. Evanston followed with the dedication of Brown and Grey Avenues, and Bridge, Payne, and Simpson Streets. Being a good neighbor, the District installed sidewalks connecting the two intersections and the bridge in 1914.

The Public Service Company of Northern Illinois (PSCNI) was allowed to withdraw water from the channel in January 1916 for use at its manufactured gas plant on Oakton Street and return the same quantity of unpolluted wastewater to the channel. However, in June 1917, a complaint was received from J.C. Sternheim of Evanston regarding the discharge of oil to the channel from the gas plant and the nuisance it caused to adjoining property owners. The District threatened to withdraw its permission to use water and PSCNI cleaned up its discharge. The manufactured gas plant, including a large gas holding tank, was built by the Northwestern Gas Light & Coke Company, predecessor of PSCNI, in 1910 and operated until the 1940s. Demolition of structures begun about 1960 was concluded by the 1970s. The District purchased the site from the Northern Illinois Gas Company (NIGC), successor to PSCNI, in stages over a period of years concluding in 1992. Site remediation began in 2013 by Nicor, successor to NIGC, to remove soils contaminated by coal tar waste. The District will use the site for its corporate purposes or may lease it for public recreation.

In subsequent years spanning two decades, the District sold excess property along the channel route, but retained sufficient land for channel maintenance. Today, most of the District-owned land along the channel is leased to municipalities and park districts for public recreational use.

Construction — General

Construction presented its own set of challenges. Similar to the North Branch improvement, there was no guidance available from the statute and no previous engineering studies upon which to base design criteria. The eight-mile long channel was divided into 11 contract sections for excavation. (See Table 1.) However, not all sections were put out for bid; employees and equipment from the channel extension at Lockport were available. Geotechnical engineering was lacking and experience was the best guide in design of the channel. The clays encountered were softer in many locations and similar to those encountered in the North Branch improvement, but dredging was not an option because of the time it would take to complete the channel. Landslides of the channel banks were a continuing problem during and after construction.

An application for a permit in September 1907 to connect the channel to Lake Michigan and to construct crib walls dutifully explained the purpose of the crib walls and the disposal of spoil from channel excavation in the lake. The permit was granted by the secretary of war in the same month authorizing the connection of the North Branch to Lake Michigan via the North Shore Channel and included: construction of the channel and rock-filled timber crib walls, limiting the total diversion of water from Lake Michigan by the District to no greater than that already authorized, requiring the work to be commenced before December 31, 1908, and completing within five years. The crib walls enclosed an area in the lake for a stilling basin to trap sand before being sucked into pumps. At this time a permit was not required by the state for creating the stilling basin or landfill in the lake.

Typical of District and construction practices of the time, the entire work was divided up among numerous contractors and the District workforce. Some of the contracts for channel excavation were limited to that alone, while others included bridge substructures and superstructures. Bridge superstructures were also divided where one contractor supplied the materials and another erected the structure. The navigation lock, Sheridan Road Bridge, and Wilmette Pumping Station were built as one integrated structure, yet involved numerous separate contracts for electrical motors, gear reduction boxes, piping, pumps, etc. In essence, the District was the general contractor performing some of the work and subcontracting numerous tasks to others.

The engineering department was busy throughout 1907 preparing railroad and road bridge designs and plans; surveying and staking the channel route, preparing excavation contract plans and specifications, and designing the crib walls. In July, several test pits were dug along the route of the channel to determine the character of the soil before construction plans were completed. The following month additional test pits at deeper depths were dug based on the results at the first set of test pits. Despite these tests, little was known of the engineering mechanics of soils until later in the century. As a result, slope failures were common and frequent during construction and for several years thereafter.

Agreements with Evanston and Wilmette prohibited disposal of excavation spoil along the route of the channel east of the Chicago & North Western Railroad, Milwaukee Division, currently known as the Metra Union Pacific / North Line. The spoil was hauled to the lake and deposited north of the cut in the bluff, creating a landfill, which presently is Gilson Park. Hauling was accomplished with dinky locomotives pulling strings of dump cars on narrow-gauge railroad tracks. To contain the deposited spoil and prevent it from drifting south and closing the cut, a rock-filled timber crib wall was built initially extending 200 feet into the lake. The crib walls were initially built beginning late in 1907 by District labor with purchased materials. At no time was concern expressed about the requirement for competitive bidding. Contractors were eventually engaged to improve the crib walls, build breakwaters, and dredge the stilling basin. (See Figure 13.)

Topsoil was stripped from the acquired properties in September between the lake and the Chicago & North Western Railroad. The topsoil could be removed using scrapers drawn by horse or mule teams and stockpiled for later use in final grading the landfill at the lakefront or for establishing a vegetative cover on channel side slopes.

Topsoil not needed by the District was later sold. Although work had already begun, it was time on September 25, 1907, for a celebration to observe the official start of work on the channel. At Central Street in Wilmette, as close to the lake as possible at 3:00 p.m., Louise Elizabeth Paullin, daughter of one of the trustees, was introduced. In a red dress with white lace trim, she moved the first shovel of dirt while the crowd of 500 cheered. Trustee Wallace G. Clark, acting as master of ceremonies in the absence of President McCormick, introduced consulting engineer Isham Randolph, who explained the history of the District, prior canal building, and the reason for the North Shore Channel. Chief engineer George Wisner next described the work to be undertaken and how the channel and bridges would be constructed. Following the ceremony, wheeled scrapers put on a show for the assembled crowd and began stripping and stockpiling the topsoil from the channel route southwest of Sheridan Road.

There was good quality sand at the lakefront that, it was believed, could be used for making concrete, and in December 1907, workers began hauling sand to various locations along the channel route where concrete would be used for bridge construction. It was believed that using free lake sand would be a savings in the cost of concrete. However, the use of lake sand was later discontinued as it was found that the uniform gradation and lack of fines rendered lake sand less desirable than limestone screenings or quarried sand. Each contractor and the District obtained their own quarried sand and crushed stone for concrete via the open market.

Besides sand, it was important to have quality cement for good concrete, and in January 1910, a contract was awarded to the Marquette Cement Manufacturing Company for 12,000 barrels of Portland cement to be used on all construction projects where needed. The delivery of Portland cement was by rail to the rail siding at Llewellyn Park in Wilmette. Each contractor was required to use Portland cement supplied by the District and responsible for hauling it from Llewellyn Park to the job site.

Coal was primarily used as fuel for the boilers in dinky locomotives, steam shovels, and other steam powered equipment. By supplying quality coal to the contractors, the District could ensure efficient performance and a reduction in soot, rather than have the contactors use cheap low-heat-value coal that would dirty the neighborhood and give rise to public complaints. A contract was awarded in June 1908 to Globe Coal Company of Chicago to furnish two types of coal. Two contracts for different types of coal were issued in 1909, but in 1910 the contractors were prepared to provide their own quality coal. Specifications were based on particle size, chemical analysis, sulfur content, heat value, and the price per ton on board at railroad sidings at either Llewellyn Park in Wilmette or Lincoln Avenue in Chicago.

After the winter shutdown, work began in March 1908 near the lakefront using machinery owned by the District, purchasing all needed materials and employing necessary labor to excavate the channel on property already acquired. Later, with cooperating lake weather, work began extending both crib walls at the lakefront. Construction of the integrated structure for the pumping station, navigation lock, and Sheridan Road Bridge began midyear with a contract for the pumps and motors. Excavation for the substructure didn't begin until December, the priority being to excavate the channel first. A contract was executed with the Chicago, Milwaukee & St. Paul Railroad for a railroad siding near Hill Street to be used for the delivery of equipment from Lockport and materials for construction. District workers laid 1,300 feet of standard gauge track for the siding and it remained the property of the District so that its use could be controlled.

1909

Construction resumed in February 1909, after the winter shutdown, excavating drainage ditches along the west edge of the channel right-of-way to collect and safely drain the surface water coming from the west. The ditch drained to the south and emptied into the North Branch, preventing flooding of the channel excavation area. East-west ditches were constructed at streets that cross the right-of-way of the channel south of Emerson Street. Main Street and Pratt Avenue were two examples of roads that didn't cross the right-of-way and where roadside ditches did extend into the right-of-way beyond the end of the road. Main Street was eventually extended to cross the channel with a bridge. Most work along the channel route was underway or had been advertised as of March. Few contracts were yet to be advertised and these were for finishing the excavation in one section, erecting three road bridges in Chicago and one in Evanston. Channel excavation by contractors generally proceeded faster than by the District.

Utility relocations or replacements were numerous in Evanston and Wilmette, where the District was responsible per the agreements with each. The only utility relocations in the city were at Foster Avenue and these were handled by the City at District expense. The work in Wilmette, under contract to William Davidson of Wilmette, involved the installation or removal of water mains, including special connections, backfilling, restoration, and landscaping. Coated bell and spigot cast iron water main with caulked and leaded joints was used. Virtually every street along the channel was involved and the work was completed in May. The work in Evanston was handled by the City, the cost thereof being reimbursed. Where water mains crossed the channel, these were suspended from the bridge and insulated. Manure was added in a jacket around the pipe in severely cold weather as an added precaution against freezing.

Due to a change in pump selection, the pumping station motor and pump contracts were cancelled and new contracts were awarded in the summer. Excavation for the substructure of the bridge-lock-pumping station began in the fall, but was delayed by a slope failure. Prior to the delay, enough excavation had been completed for foundation piles to be driven and forms built for the pumping station foundation slab.

1910

In April, the chief engineer estimated that with good weather the construction work would be completed by the end of the year. That would allow sewage to be turned into the channel. Wilmette was expected to be ready to divert their sewers, but Evanston would take longer to complete their sewer work. In July, Evanston announced plans for the construction of a sewer to discharge to the lake, but the District objected. Since the channel would be completed before the sewer construction, the sewer was redesigned to discharge to the channel. North of Wilmette, no action was underway by other suburbs to connect their sewers to the channel.

To prevent overflow and erosion of the channel bank, storm drainage in roadside ditches needed to be connected to the channel. In August 1910, a contract was awarded to Carden-Callahan Company of Chicago for construction of ditch outfalls at numerous locations where streets either abutted or crossed the channel. Between Emerson Street in Evanston and Bryn Mawr Avenue in Chicago, 30 sewer outfalls were built. Two designs were used, and the standard outfalls were designed for a low water level in the channel of 2.5 feet below CCD. To prevent channel side slope erosion at other locations where the ditch invert was higher, the outfall was built with a sloping concrete chute embedded in the channel side slope. Carden completed this work at all locations by November.

Concrete outfalls and slope paving were completed for the Dewey Avenue and Emerson Street sewers that were cut during excavation of the channel. The new outlets would allow the flow to drop to the bottom of the excavation without causing erosion of the channel side slope. During channel excavation, those sewers drained into the excavation; the sewage was then pumped into downstream sewers that drained to Lake Michigan. A result was excavation work going on in the presence of sewage.

With channel excavation mostly completed, priority shifted to construction of the pumping station and lock. In the space of eight months the following were completed: the lock chamber and two sets of miter gates, a pumping station and all electrical and mechanical equipment, and four reinforced concrete retaining walls on the lake-ward and land-ward sides. As a result, water filled the excavated channel from Sheridan Road to Howard Street by the end of November. Structural steel superstructure framing for the bridge was well along by the end of the year.

Side slope failures were common during excavation of the channel. On November 29, before subfreezing weather arrived, water was turned into the channel to reduce the risk of side slope failures. Most side slope failures had occurred between Sheridan Road and Main Street because the soil underlying this reach was too soft to sustain the stockpiling of spoil. Residual channel excavation, not totally complete, was left to be completed the following year using dipper dredges, the spoil to be removed by scow.

Like the groundbreaking ceremony just over three years earlier, a celebration was staged. On November 29, a crowd of 1,000, complete with school children, gathered on the slopes bordering the stilling basin. They watched laborers hand dig the soil blocking the entrance to a wooden flume embedded into the embankment separating the pumping station from the stilling basin. As the laborers got close to the last few shovels of soil, President McCormick and Trustee Baker stepped forward at 10:30 a.m. in the chilly breeze off the lake, took hold of the shovels and completed the cut. Water began to flow into the channel. The children began yelling, a few dinky locomotives began blowing their whistles, and the adults clapped. There was no brass band or speeches and the event wasn't acknowledged in the proceedings. As the crowd dissipated, water slowly filled four miles of the channel.

The timing of the celebration suited the political calendar; President McCormick, having completed his five-year term and losing his bid for reelection, was out of office a week later.

1911 AND BEYOND

Despite filling half the length of the channel, the work was far from complete. The various contractors at the pumping station and lock completed their tasks in 1911 and continuous pumping of lake water for dilution began in April. In subsequent years, the main focus of the District was on controlling side slope failure, a continuing problem (discussed later) that would render the channel of little use for conveying sewage from North Shore communities and dilution water from Lake Michigan.

In June 1911, the District constructed an outfall in Darrow Avenue for the Evanston sewer because the sewer was blocked during channel construction. The crib walls for the stilling basin needed reinforcing; also needed was a breakwater to withstand the force of storms on Lake Michigan. In July 1911, a permit was obtained for modifications to reduce the accumulation of sand in the stilling basin from storms. The depth of the basin was 12 feet below CCD, but periodic dredging would still be required to maintain this depth. While work on the basin wasn't completed until 1922, small boats began to moor at what soon became known as Wilmette Harbor.

Meanwhile, in the latter part of 1911, a tempest was brewing in the North Shore suburbs. Sewage, still being discharged to the lake, was polluting local water supplies. Appeals to the District trustees for relief were turned aside. Some trustees asserted that the municipalities should build intercepting sewers to the new channel, while the municipalities, pointing to the taxes paid to the District, asserted this was the District's responsibility. Villages along the North Branch threatened to sue the District for relief from sewage being discharged to the river. The resolution is explained in the next chapter.

For the casual reader the foregoing is perhaps sufficient to understand the history of construction of the North Shore Channel. However, the armchair mechanic, engineer, or local historian may want to read on to learn the salient details of this endeavor.

Wilmette Stilling Basin and Harbor

The first construction work on the North Shore Channel, to create an inlet for lake water to be pumped into the channel, occurred at the lakefront in Wilmette. Based on experience with other lakefront structures, particularly the Thirty-Ninth Street Pumping Station, it was known that the littoral drift of sand could block such an inlet and the sand could easily be sucked into the pumps. Thus, today's Wilmette Harbor was originally created as a stilling basin to allow sand to settle. The pumping station intake was set back about 200 feet from the 1895 shoreline, but construction of the landfill and stilling basin positioned the intake approximately 1,000 feet from the stilling basin opening to the lake. Rock-filled timber crib walls were extended into the lake from the shoreline. Later, to create the stilling basin, concrete retaining walls were extended on either side of the intake and connected to the crib walls. In addition, the crib wall on the north side of the stilling basin also served as the south edge of the landfill formed from channel excavation spoil. (See Figure 13.)

The initial construction of the crib walls and concrete retaining walls was accomplished by District labor under the guidance of the engineering department. Later, contracts were awarded for dredging the stilling basin, for reinforcing the rock-filled timber crib walls, and for other work associated with the stilling basin. In April 1907, two contracts to furnish and deliver lumber were awarded: the Mears-Slayton Lumber Company to supply yellow pine, and the Chicago Car Lumber Company to supply hemlock. Lumber of various sizes and lengths was delivered to the Llewellyn Park railroad siding. Two scows were purchased, one for use by the carpenters building the crib wall and the other for hauling rock to the work site. The rock was towed by hired tug boats from the spoil piles along the Sanitary & Ship Canal near Lemont.

In mid-September 1907, two weeks prior to the formal ground breaking ceremony, construction began for the crib wall in the lake; also, excavation began for the foundation of the concrete retaining walls lakeward of the pumping station. An embankment was built to isolate the excavation area from the lake. After three months of work on crib wall construction, it was getting late in the season and storms with wind-whipped waves were causing frequent delays. However, the first 200-foot section of crib wall number one was completed; placing spoil in the lake could begin.

In May 1908, after a long winter shutdown, work resumed and continued through the year with few weather-caused interruptions. The crib walls were 12 or 16 feet deep, the latter used in deeper water. Some excavation of the lake bed was performed to position the crib wall on a relatively level surface. The crib wall, partially filled with rock, would sink and set itself in the lake bottom. Track was laid on the crib for dumping rock into the crib wall; on the northwest crib wall the track was also used for dumping spoil into the landfill. The crib walls were filled with rock to above the water line. A stiffleg derrick on a scow was used to unload and place the rock, which was supplied in dump cars or on another scow. The scows were maneuvered by the harbor tug *Princess*, leased by the District. Occasionally, large waves from strong storms would wash out newly constructed crib walls. By the end of the year, over 1,800 feet of crib wall were completed, enclosing the stilling basin except for the opening to the lake.

Over the winter, extensive damage to the crib walls was caused by strong waves, and plans for repair and reinforcement were prepared in the spring of 1909. Rock from the lake bed and stilling basin dislodged by the winter storms was retrieved and placed back in the crib walls. In July, a contract was awarded to Great Lakes to construct a short breakwater—a rock-filled timber crib wall to narrow the stilling basin inlet and ward off waves. In addition, Great Lakes dredged the stilling basin inlet channel and reinforced the crib walls by installing pilings and tie-rods and attaching oak whalers. Hydraulic dredging began in the inlet channel, but ceased after three days due to the number and size of large rocks encountered. The dredging, which lasted to early

October because of frequent interruptions due to rough lake conditions, was completed by a dipper dredge. Winter weather forced Great Lakes to shut down for the year with about half of the work completed. Work resumed in June 1910 and was completed by early August.

More dredging of the stilling basin inlet was needed but wasn't urgent; in March 1911, that work, combined with a dredging contract for the North Branch, was awarded to Great Lakes. Clay needed to be dredged from the stilling basin inlet channel for adequate flow to the pumping station. As in the case of the earlier dredging of clay, the spoil was placed in the landfill. The dredging was completed in June.

Sand bars blocking the inlet channel continued to be a problem. In May 1916, a breakwater was planned north of the inlet channel. In September 1917, a sunken scow outside the entrance channel was removed; it had been in place for several years and was causing the formation of a permanent sand bar. In 1920, construction of a 500-foot-long rock-filled timber crib breakwater began with the award of a contract to Great Lakes. The contract also included placing concrete caps on structurally sound crib walls and removing structurally unsound crib walls, replacing them with reinforced concrete walls founded on pilings. Once again, dredging the stilling basin was necessary. All work was completed by 1922.

Up until this time, the stilling basin had been used frequently for construction by the District, discouraging the mooring of private boats. However, as time went by, more boat owners began using the stilling basin as a harbor.

Wilmette Pumping Station, Navigation Lock, and Sheridan Road Bridge

Three functional elements near the lakefront—the navigation lock, pumping station, and Sheridan Road Bridge—were integrated into one structure. The initial construction of these elements was accomplished by District labor under the guidance of the engineering department. Later, contracts were awarded for different parts of the project: the electrical and mechanical equipment for the pumping station and lock, the metal work and other finishing details for the pumping station, ornamental stone facing for the bridge, and paving for Sheridan Road. Spoil from the excavation of the site was loaded into dump cars and towed by dinky locomotives to the northwest side of crib wall number one, where it was deposited in the landfill.

Early planning contemplated a coal-fired, steam-powered pumping station near the north end of the channel between the lakefront and the Chicago & North Western Railroad crossing. However, the eventual agreements with Evanston and Wilmette precluded coal-fired facilities and required electrically driven pumps. Next considered was a coal-fired electrical generating station outside the municipal jurisdictions and transmission of the electricity to the pumping station, but hydroelectric power from the Lockport powerhouse became available, and it was chosen. To operate a pumping station, electrically driven pumps were known to cost less than coal-fired steam-driven pumps, even including the cost of a transmission line from the main substation in the city to Wilmette. By July 1907, it was decided to locate the pumping station at the lakefront so there would be no channel lake ward of the station for receiving discharge from sewers.

For the rest of 1907 and early 1908, the engineering department prepared the design, as an integrated structure, of the pumping station, navigation lock, and Sheridan Road Bridge, including associated channel walls and roadway approaches. In July 1908, a contract was awarded to the Camden Iron Works of Camden, New Jersey (Camden), for the centrifugal pumps and a high pressure pumping set for station water. The contract for electric motors was awarded to the Western Electric Company of Chicago. Camden was required to manufacture, deliver, and install eight centrifugal pumps and a high pressure pump and motor set, which were to be used for pump priming and other station service needs such as cleaning and irrigation. To drive the pumps, Western Electric was required to manufacture, deliver, and install two alternating current induction type electric motors. The plan was to have gear trains or belts and pulleys transmit the motor power to the pumping units.

In December 1908, a steam shovel began excavation for the foundation of the pumping station west of the Sheridan Road temporary trestle. An earth embankment along the shoreline between the crib walls kept lake water out of the foundation site. Little was accomplished over the winter, and in 1909 excavation proceeded slowly because the District was concentrating on nearby channel excavation. Spoil was hauled by rail and deposited in the landfill. Timber piles were driven for the pumping station foundation. By October carpenters were building forms for the concrete foundation slab. In November a slope failure filled part of the excavation and stopped work. Little else was accomplished before the winter shutdown.

Meanwhile, by March 1909, Camden was not making progress in manufacturing the centrifugal pumps, and the engineering department was considering the use of screw pumps as an alternative. Axial flow screw pumps, just as efficient, appeared more practical for the low head pumping and physical configuration. In June, the contract with Camden was cancelled and a contract was awarded to Allis-Chalmers of Chicago (Allis-Chalmers) for four screw pumps. Different electrical motors were needed and, by fortunate coincidence, Western Electric had requested a substitution on another unrelated contract so the District was able to cancel the contract for the electrical motors in exchange for approval of the substitution. For the new electrical motors, a contract was awarded to W.A. Jackson Company of Chicago (W.A. Jackson). In February 1910, the contract for the high pressure pumping unit was awarded to Allis-Chalmers.

The decision to locate the pumping station at the lakefront and combine it with the Sheridan Road Bridge raised concern about the appearance of the structure. The architectural design of the structure was based on sketches by President McCormick, who was impressed by beautiful bridges on his European travels. In August 1909, Bedford stone, rather than concrete, was selected for the exterior; this betterment would achieve a more handsome appearance. The stone was purchased from an Indiana quarry and allowed to season for one year. The next month a contract was awarded to Adam Groth & Company of Joliet. Stone was delivered to Groth's Joliet plant where it would age and then be cut. In June 1910, the cutting of the stone was completed; the cut stone was held in Joliet where it awaited delivery and installation.

In February 1910, the Milwaukee Bridge Company of Milwaukee, Wisconsin (Milwaukee Bridge), was awarded a contract for fabrication, delivery, and erection of metalwork for the pumping station and bridge. In March, to clear the area for construction of the stilling basin's northeast and northwest concrete retaining walls, the Sheridan Road temporary trestle was removed and traffic to cross the channel construction area was re-routed via Linden Avenue. In May, a contract was awarded to the National Contracting Company of Chicago (National Contracting) for fabrication and erection of machinery, lock gates, and collateral work for the pumping station. For the main pumps, National Contracting furnished and installed four gear reduction units, connected the pumps and motors, and tested the motors with the gear reduction unit. For the lock gates, National Contracting furnished and erected two pairs of lock gates complete with their miter sills and operating machinery. For the pumping station, National did the following: it furnished, delivered, and installed intake screens, stop logs, winches, and chains for four flap gates; it furnished a tripod mounted hand winch for the stop logs and screens; and it furnished and installed a high pressure piping system. Basically, National Contracting was required to get the pumping station and lock operational.

The following month, using a steam shovel, excavation resumed just west of the station. To keep the excavation dry for the foundations of the retaining walls, pumps were run continuously. Lighting was set up, and night work allowed the District to complete the excavation and build forms for the complex pump tunnels. Although 63% of the excavation was complete, there was no time to waste if the pumping station was to be finished by the end of November. Due to continual wet weather and a large storm at the end of April, the foundation excavation was flooded all of May. A dragline was moved from Section 3 for use in placing concrete. Toward the end of the month the carpenters were building the pump tunnel forms. In June, following the driving of foundation piles and working double shift, excavation was completed and forms built for the concrete retaining walls and lock chamber foundation. Concrete pours were begun on the southwest wall. A slope failure of the east channel bank caused suspension of work for a week to clear the clay in the channel bottom.

Better weather during the summer allowed much progress. All excavation was completed, Wakefield piling was driven around the site, the last of 1,401 foundation piles were driven, and concrete pours for the pump tunnels and the walls for the pumping station and lock were completed. Outside the confines of the pumping station and lock, work progressed on the concrete retaining walls. First to be completed were the southwest concrete retaining wall and sewer outlet, followed by the southeast wall, the downstream lock chamber and channel floors, and the lock chamber wall extension. Even another side slope failure early in August caused little delay; the clay was quickly removed. Last to be completed were the northeast and northwest retaining walls enclosing the stilling basin between the pumping station and former shoreline; by the end of October, those walls were tied into the shoreline ends of the rock-filled crib walls.

In August 1910, a contract was awarded to Henry Gilsdorff & Sons of Chicago (Gilsdorff) for all labor and materials for masonry, concrete, water proofing, tile work, lathing, plastering, mill work, plumbing, sewerage, glazing, architectural metal work, and painting. In the fall of 1910 the site was a busy place with all the contractors working to finish their tasks: Jackson delivering the pump motors in October; National Contracting installing the pump discharge flap gates, working on the lock gates and pump machinery; Milwaukee Bridge setting steel columns and roof framing; Allis-Chalmers installing the screw pumps and bearings; W.A. Jackson installing the electrical motors; and Groth fitting the cut stone facing to exterior walls. The site was not proper for National Contracting to conduct an efficiency test on the electrical motor and gear reduction unit, so one set was transported to Hanna Engineering Works in the city. There, the efficiency was determined to be 88.5%; following the test, the equipment was returned to and installed in the pumping station.

By November, backfill was placed behind the several completed walls; the abutments at both ends of the Sheridan Road Bridge were completed and Gilsdorff was busy with myriad finishing tasks for doors, windows, plastering, painting, and roadway curbs. Gilsdorff formed and poured the reinforced Portland cement concrete roof of the pumping station/bridge structure, but the chief engineer determined that concrete would not be suitable for the roadway surface. A contract was awarded to Thomas W. Kelly to pave the roadway, using asphalt concrete on the structure and a macadam pavement for the approaches. As an additional precaution, horizontal concrete struts were built between the walls beneath the channel bottom; the purpose was to protect the walls on the channel side of the pumping station from being pushed inward by the lateral force of the soft clays.

Early in 1911, National Contracting finished the erection of machinery, lock gates, and collateral work at the pumping station and their final payment was held pending tests of equipment. National Contracting had been delayed in completing their work for several reasons: the District had not completed the structure and had not installed electrical wiring; also, the District had issued a change order for installation of electrical motors and operators for the lock gate machinery. The lock gates were originally designed and built to be operated manually, but such operation proved to be slow, consuming considerable time and labor.

To memorialize the beautiful and useful integrated structure, four bronze plaques were designed and mounted on the Sheridan Road Bridge. The four plaques, still in place, display a 1910 map of the District and list the members of the board of trustees; McCormick's name is listed twice, as president and as a member. The plaques have the following credit for the structure: *The Sanitary District of Chicago, North Shore Channel, Sheridan Road Bridge and Wilmette Pumping Station, Designed from President Robert R. McCormick's Original Sketch by the Engineering Department of the Sanitary District of Chicago, under the Direction of George Wisner, Chief Engineer, Carlton R. Dart, Bridge Engineer, and Frederick L. Barrett, Architect.* The plaques also list 19 members of the engineering department; one of the names is E.J. Kelly, Assistant Engineer, later to be the City's mayor.

To determine if Allis-Chalmers met the specifications, the efficiency of the screw pumps was measured in place. Two pools were created—a smaller upstream inlet pool east of the pumping station and a larger downstream discharge pool. The inlet pool was confined by the northeast and northwest retaining walls, the pumping station and lock chamber, and the earthen embankment between crib walls number one and number two. The discharge pool was confined in the channel between the pumping station and an earth dam across the channel at Howard Street. A weir was built in the lock chamber providing a means to measure the rate of flow between the two pools. A wooden flume in the earthen embankment was used to allow water to flow at a controlled rate into the excavated channel. During the ceremony on November 29, 1910, described earlier, at 10:40 a.m. water was let in slowly; the water level in the channel north of the earth dam at Howard Street slowly rose, becoming equal with the lake level by midnight.

The testing of the pumps occurred in 1911, with preliminary tests in January and the final test in March. As a check on the weir, District engineers ran 14 tests using the rise in water level of the lower pool and the calculation of volumetric rate of flow. The flow determined by the volumetric method agreed closely with the weir measurements. For a further check on the weir, the flow in the channel was measured by a current meter. Finally, on March 24, 1911, the official test of pump number 2 was made, but the pump manufacturer declined to witness the test due to a discrepancy in the preliminary weir and current meter flow measurements. The official test result showed a pump efficiency of 61%; however, the contract guaranteed pump efficiency of 67.5% with a penalty or bonus for each 0.25% below or above the guaranteed efficiency. Each pump was rated at 250 cfs, and with four pumps in operation, half of the discharge capacity of the channel could be delivered to comply with the statutory dilution requirement.

Following the official test, the manufacturer raised many questions about the accuracy of the measurements. Responses were given that verified the accuracy of the measurements. An investigation was also made of the current meter measurements and, using a directional vane along with the current meters, it was found that the flow in the measurement cross-section was not uniform. At the bottom of the channel the flow was parallel to the length of the channel, but at higher elevations the flow was transverse, indicating the presence of eddy currents. When the current meter velocity measurements were corrected for flow direction, the difference between the flow rate determined by current meters and the weir was 0.4%. The determination of pump efficiency was sustained; the final payment to Allis-Chalmers was reduced by the specified penalty.

After completion and verification of the pump test results, the dam at Howard Street was removed. Now the channel was open from Sheridan Road to Lawrence Avenue. In April, cofferdams were installed, the four pump tunnels were dewatered, then cleaned, then flooded again, and the cofferdams removed. On April 18, pumping began; pump 2 was intermittent at first until the flume in the lakeside dam was enlarged. Thereafter, pumping was continuous with at least one pump, alternating pumps 1, 2, and 3. Pump 4 was being worked on for noise and vibration problems. In May, the lakeside earthen embankment and lock chamber weir were removed.

In February 1911, Milwaukee Bridge finished the last of the steel framing; its start had been delayed by the District not having the foundations completed. Gilsdorff finished in July, considerably beyond the contract date, due to the numerous contractors on the site. Groth placed the last of the Bedford stone on the bridge in June. In August, a contract was awarded to Demling & Wendt of Chicago to install sidewalks and drains along Sheridan Road; this work was completed in November. Another contract, at the request of the Village of Wilmette, was awarded to Thomas W. Kelly for pavement, curbs, and drainage work on the south approach. Changes were made in the operation of the flap gates on the pump discharge conduits. Finally, in November, electricians installed 30 street light poles and lamps on the Sheridan Road Bridge and approaches for safety.

Closeout of the many contracts waited until 1912, after all work was completed. All the contractors had exceeded their completion dates, but due to delays by the District that restricted access and crowded working conditions, none were penalized for late completion. In November, an agreement with Wilmette was approved for the use of 141 square feet of space at the pumping station for the village water supply system. The village paid for the electricity used by pumps and other equipment and the District obtained water from the village at no cost.

In 1913, a public comfort station was installed adjacent to the pumping station under the northwest span of the bridge for the convenience of District employees and the public using the harbor and lakefront park; concrete stairways were built from the roadway level to the pumping station level. Max Lundguth & Company of Chicago built the sidewalks and stairways, and the Van Etten Brothers built the public comfort station and a storeroom. In September 1915, a diver was employed to remove deposits on the screens and to inspect the bearings on the shaft of the turbines. The inspection revealed the need to replace the bearings; the work was completed by the end of the year. In November 1916, a 150-horsepower motor rotor that had burned out was rewound.

Pumping lake water into the channel for dilution of sewage was continuous from 1911 through 1938. During this period there was no control structure between Lake Michigan and the Chicago River. On January 1, 1939, the Chicago River Controlling Works went into operation, and with the new control on the Chicago River and per direction from the Corps, the District began to manipulate water levels in the canal system at a lower level. Given variations in lake levels, maintaining the canal system at a lower water level than Lake Michigan was a positive means to prevent discharge to the lake. During storm periods that lower control level also helped to improve the outfall capacity of the local sewers discharging to the canal. Thus, the North Shore Channel water level was lower; and when the lake was sufficiently high, water could be diverted by gravity at Wilmette, saving the cost of electrical energy for pump operation. However, when the lake level was too low for gravity diversion, pumping continued. Even with collection and treatment of sewage, it was necessary to divert lake water to reduce the incidence of odors in the North Shore Channel and North Branch because of frequent overflow of combined sewers caused by storms.

Channel Excavation Section by Section

As mentioned earlier, the channel was built in 11 sections but the various sections were vastly different, as were the demands placed on the contractors. Three of the sections were built by District labor, including sections originally designated as 10 and 11, which were combined as one section. (See Table 1.) Except at the three railroad bridges, the channel was trapezoidal in shape, with a bottom width that varied from 26 to 50 feet. The bottom width was 50 feet from Lawrence Avenue to the North Branch Dam, 30 feet from the dam to Foster Avenue, 26 feet from Foster Avenue to Howard Street, 38 feet from Howard Street to Main Street, and 30 feet from Main Street to Sheridan Road. The bottom elevation was 12 feet below CCD, with no gradient, from Lawrence Avenue to the North Branch Dam. There was a vertical drop to 12.7 feet below CCD at the dam, and a rising gradient of 0.00005 to a bottom elevation of 10.69 feet below CCD at Sheridan Road. Extra width was needed in the reach from Howard Street to Main Street because the bottom elevation was raised 1.6 feet due to the skewed railroad bridge south of Oakton Street and a major manufactured gas main in a tunnel under the channel at Oakton Street. The drop of 0.7 feet in the invert near the dam is unexplained and has no hydraulic significance; it would soon fill with sediment.

SECTION 1

Section 1, a 3,250-foot-long channel, was constructed by District labor. The pumping station, navigation lock, part of the stilling basin, and attendant walls occupied the first 600 feet from the lake shoreline. During the last three months of 1907, wheeled scrapers removed topsoil from Laurel Avenue to Sheridan Road, but once stiff clays were encountered, the scrapers weren't effective, and in late December work was shut down for the winter. Substructure work for four road bridges—Sheridan Road, Linden Avenue, and Hill and Isabella Streets—had already begun.

Initially, the trustees wanted District labor for the early work near the lakefront because they believed excavation of the channel east of Dewey Avenue in Evanston would take a year longer than the channel west and south of that point. Extra time would be needed to move the spoil and deposit it at the lakefront rather than beside the channel. In addition, the crib wall needed to be extended as the spoil was deposited in the lake; village sewer and water pipes had to be diverted and/or relocated, and numerous road bridges, two railroad bridges, and a pumping station had to be built. A normal construction contract would have to wait until all land was acquired and all design details worked out. This complexity, the trustees believed, could not be reduced to specifications for one or more contracts; it was assumed contractors would charge unreasonably high fees for such uncertainty.

In retrospect, the chief engineer did not recommend this course of action. The District had previously issued contracts that included contingencies and uncertainties, and competitive bids had been obtained. Perhaps the trustees had in mind that there was equipment available that had been confiscated from a forfeited contractor on a Lockport project; perhaps they wanted to add employees to the payroll. The thinking was never made explicit in the public record. As it turned out, work near the lakefront was more complicated and took longer than work done by contract at other locations.

Over the winter, District crews laid 4,800 feet of narrow-gauge track for the steam shovels, dinky locomotives, and dump cars. In April 1908, excavation began at the lakefront bluff, and the trestles and approach roadways for the temporary Sheridan Road and Hill Street diversion bridges were completed. Not all work was directly for the channel; repairs were made to the District field office on Grant Street and to the building on Noyes Street used to house Austrian laborers. Rain in May caused excavation delays and slumping of the spoil piles, wooden flumes were built to divert local sewers, and the District field office was moved from Grant Street in Evanston to Llewellyn Park in Wilmette. Other projects were completed over the summer: the vacated building on Grant Street was converted into a bunkhouse for more laborers, a steam shovel began channel excavation in August, and a three-day delay occurred due to a strike of laborers seeking a wage increase, which was denied. Late in the year, though excavation progress slowed due to soft, sticky clay being difficult to handle, a roundhouse was built for the dinky locomotives and excavation began for bridge abutments. The winter shutdown occurred in late December.

In 1909, excavation started late due to cold and wet weather. Repair work was performed on equipment; tracks were replaced where slope failures occurred and where rails from the landfill were washed into the lake by storms. A steam shovel was excavating the channel near Hill Street, working its way south to Isabella Street in early April, but in late April a storm flooded the excavation, dislodged tracks, and damaged a sewer outlet. The earth plug at the lake front was breached to let the stormwater out of the excavation, and local roads were repaired from the storm damage. In May, the excavation reached the end of the section near Jenks Street; on good days, the one steam shovel kept six dinky locomotives busy hauling loaded dump cars to the lakefront landfill. For the balance of the year the steam shovel worked back and forth along the channel excavation, widening the cut to the specified width. Track crews were constantly moving and shifting track to keep up with the excavation progress, but a storm in July brought work to a standstill. Again the excavation was flooded, and the stormwater flowed out to the lake for three days. Work resumed after pumping the water out, and by the end of the year excavation was 80% complete. No excavation was performed in December due to an early winter shutdown.

With an earlier spring in March 1910, excavation work resumed. Where the cut was completed to full width, laborers were finishing the channel side slopes by hand. In June, some laborers were shifted to excavation work at the pumping station near the lakefront. For the balance of the summer and into fall, until late November, work continued day and night, completing the full width cut and, when channel excavation was completed, finishing the side slopes. The Laurel Avenue sewer outlet was built and the channel was ready for water. Equipment was moved to the Llewellyn Park railroad siding for storage.

SECTION 2

Section 2, a 4,080-foot-long channel, was also constructed by District labor. This section included two railroad crossings—at the Chicago, Milwaukee & St. Paul Railroad, located north of Central Street, presently the Chicago Transit Authority Purple Line, and at the Chicago & North Western Railroad, presently the Metra Union Pacific / North Line, and two road crossings—at Central and Lincoln Streets in Evanston. Substructure work for railroad and road bridges was included and performed by District labor.

In May 1908, a steam shovel was set up for excavation, but little channel excavation was completed that year as the focus was on Section 1. However, bridge substructure work did proceed. The next year began with excavation proceeding from the north end near the Chicago, Milwaukee & St. Paul Railroad crossing, but time was lost due to a railroad embankment slope failure and cold weather. When work was progressing, seven dinky locomotives were kept busy hauling excavated channel spoil that was used as fill for the temporary embankment for the railroad diversion around the permanent bridge construction site. Four dinky locomotives were used to haul spoil not needed for the embankment to the lakefront. Lingering cold weather intermittently delayed work. In late April 1909, a flood occurred, completely filling the excavation and inundating one steam shovel and one dinky locomotive. Evanston rerouted the Asbury Street sewer and the police and fire communication lines. A drainage ditch was excavated along the west right-of-way to intercept surface runoff from the neighboring golf course.

By August, the initial cut had reached the southwestern end of the section and, except for a few rain delays, excavation work proceeded expeditiously; two steam shovels were used, necessitating the purchase of five new dinky locomotives. Excavation was more than half completed by the end of December and the winter shutdown. In April the following year, work finally resumed. Except for the tight working conditions at the two railroad bridge construction sites, the dirt was literally flying to the lakefront landfill with day and night shifts. Manual finishing of the channel side slopes began in September; excavation was completed by late November.

SECTION 3

Section 3 was 1,480 feet in length and had only one road crossing. The east section limit was between the Chicago & North Western Railroad and West Railroad Avenue as shown on Figure 10, resulting in the temporary diversion for the railroad in Section 2 to be in Section 3. In March 1909, prior to the award of a contract for Section 3, District labor completed the road diversion to the west of West Railroad Avenue. The Chicago & North Western Railroad began constructing the track roadbed diversion and temporary bridge trestle construction, West Railroad Avenue was closed, and traffic was rerouted via Simpson, Darrow, and Grant Streets.

In August 1909, a contract was awarded to Schnable & Quinn of Chicago. In addition to the excavation of the channel, the contractor was required to do the following: maintain ditches and levees for surface drainage, construct West Railroad Avenue bridge abutments and piers, erect the bridge superstructure, and remove the temporary road diversion. In the remaining working months of 1909, Schnable & Quinn began the erection of their dragline excavator and removed and stockpiled topsoil from the channel area. Some of the topsoil was removed from the site by the Peterson Nursery in Chicago.

Early in 1910, the contractor began channel excavation east of Darrow Avenue and was rerouting sewers north and south of the channel route. The dragline excavator, more efficient for trapezoidal-shaped open channel excavation than the steam shovel, proceeded much faster than in Sections 1 and 2. By May the contractor was excavating west of West Railroad Avenue and was finishing the south slope. Per the agreement with Evanston, spoil was placed south of the channel and west of Dewey Avenue.

South of the channel, the Northwestern Gas Light & Coke Company installed a gas main along Brown Avenue and connected it on a diagonal across the channel right-of-way to a gas main north of the channel in Grey Street. In July, with the new main in service, the old gas main in Dewey Street was removed and excavation proceeded. Schnable & Quinn constructed a wooden flume to convey sewage across the channel at Dewey, a major south-flowing sewer that couldn't be rerouted. By summer, excavation near West Railroad Avenue and the flume were completed, and the contractor, working day and night, moved the dragline to the west end of the section. The road diversion at West Railroad Avenue was opened to traffic, and Darrow Street was closed so that excavation at the west end of the section could be completed. The north and south slopes were completed between Dodge Street and Darrow Avenue; the contract was 75% complete.

In August, two side slope failures occurred on the north side of the channel, one failure completely destroying the wooden sewer flume and causing sewage to flow into the excavation. With the contractor working day and night, the destroyed flume was removed and sewage in the channel was pumped into the Dewey Street sewer south of the channel. All excavation was completed by the end of September. More slope failures occurred in October and excavation continued to maintain the specified channel width. All excavation was completed in November, and the District had constructed an outfall for the Dewey Avenue sewer by the end of the month.

Although completion was much later than called for, most of the delay was beyond the control of the contractor. The District purchased the dragline and other equipment from Schnable & Quinn for work on Section 7.

SECTION 4

Section 4 included three road crossings in a 5,940-foot-long section: at Bridge, Emerson, and Church Streets. In June 1908, having acquired sufficient property between Church and Dodge, the contract was awarded to James O. Heyworth Company of Chicago (Heyworth). In addition to excavating the channel, Heyworth was required to alter and/or extend sewer, water, and public utilities; construct bridge abutments and concrete pier foundations; erect structural steel and complete bridges at the three crossings; and furnish, maintain, and remove temporary bridges at Church and Emerson Streets. This channel excavation contract and others didn't specify the slope of the channel bank; the contractor was responsible to protect the completed excavation from erosion, slides, and upheavals. *Upheaval* refers to soil expansion when overburden was removed. The spoil deposited in designated areas was to be graded to slopes not steeper than one vertical on one-and-one-half horizontal, and the toe of the spoil pile slope was to be no closer than 25 feet from the top of the channel bank. In July, Heyworth's offer to purchase and remove topsoil was accepted.

Heyworth began work immediately, working through the summer to excavate the channel and a collateral ditch for drainage and initiate bridge substructure work. A second dragline excavator was added in September, the contractor working day and night for the rest of the year from the west side of the channel south of Emerson Street. In October, Heyworth brought in a grading machine to begin finishing the channel side slopes. The dragline continued into Section 5 since Heyworth had also been awarded that contract. By fall, bridge foundation work was completed. By the end of the year, channel excavation continued day and night and all the bridge superstructure steel had been delivered and erection begun. The contract was 40% complete in just seven months.

In January 1909, Heyworth began excavation by dragline in the vicinity of Emerson Street, working northeast along the northwest channel bank, and began the temporary bridge at Emerson Street. Frequent delays were caused by the cold, wet winter weather. By spring, the dragline reached Bridge Street and was moved to the opposite side of the channel, where excavation began along the southeast side. A new excavator began at Church Street, working to the south on the east side of the channel. The new dragline, named *McCormick,* was capable of finishing the slopes as it progressed. Working day and night with the two draglines, Heyworth was 75% complete by the end of July. An August storm caused some delays, and drainage had to be pumped to the sewer in Grey Avenue since the collateral ditch for drainage to the south wasn't yet completed. Working from both sides of the channel, *McCormick* and *Evanston*, another named dragline, were making good progress. There were a couple of side slope failures, but they were cleaned out and the channel bank was made less steep. The excavation work reached 98% by the end of 1909, including the entire length of the collateral drainage ditch.

After the first of the year, a slope failure near Church Street was cleaned out with a dragline. All excavation was now complete with two exceptions: minor finishing, and clearing of ditches along Church and Emerson Streets that had become clogged with debris over the winter.

SECTION 5

Construction of Section 5, slightly less than one-half mile in length, was awarded to Heyworth late in 1908. It required the same work as the Section 4, except for having only one bridge at Dempster Street. Some excavation work began immediately, but then shut down for the winter. In May 1909, Heyworth resumed excavation with the dragline *Evanston,* starting at Lake Street and working southward on the east side of the channel. With day and night shifts, work progressed rapidly, achieving 53% completion by the end of July, including the Dempster Street bridge abutments and foundations. In August, the collateral drainage ditch was completed, except for the connection to the Section 4 collateral ditch. In October, a slope failure occurred on the east bank near Crain Street, and a large amount of soil slid into the channel. Some of the material was removed and, with about 96% of the work completed, work wound down for winter. Early in 1910, Heyworth used a dragline to clear two areas, the remainder of the material in the bottom of the channel near Lake Street and another slope failure near Dempster Street on the west bank. All work was completed by September.

Heyworth claimed extra compensation for both Sections 4 and 5 for removal of soil resulting from slope failures; the District counter-claimed a reduction for soil remaining in the channel. Testimony in court revealed the engineering department's lack of knowledge of the soft clay stratum underlying the firm clay. The slope failures were precipitated by the oozing of the soft stratum into the excavated channel prism from under the east side slope. Removal of the soil in the channel, as ordered by the District, only exacerbated the unstable soil condition and allowed more soil to slide into the void. In April 1911, a settlement was reached that compensated Heyworth for removal of the extra soil.

SECTION 6

For Section 6, a short 1,350-foot section, land acquisition south of Main Street was delayed, but in the interest of progress a contract was awarded to John T. Fanning Company of Chicago (Fanning). Fanning was required only to excavate the channel and construct ditches and levees to handle surface drainage. Fanning began work in August, erecting a cableway excavator that was ready for use in October. Once started, excavation continued through the end of the year. A cableway excavator allowed excavation to proceed on both sides of the channel simultaneously. Some excavation was accomplished in January, but excess water became a problem as soon as the frozen ground thawed. As excavation of the west drainage ditch resumed, pumping water out of the cut occupied much of March. Excavation was interrupted by flooding in late April and pumping occupied the early part of May.

By late July, when the west tower of the excavator was destroyed by fire, Fanning was 55% complete. Work came to a stop in August while the tower was rebuilt. Excavation resumed and proceeded well until October, when two slope failures occurred on the east bank north of Main Street; another failure occurred in December. The contract stood at 76% complete at the end of the year. In February, Fanning started by dismantling the cableway excavator and assembled a small dragline to finish out his contract. By working day and night, Fanning was able to make up for lost time; the section reached 78% completion by April. Final grading of the channel banks and removal of the slope failure material in the channel bottom proceeded when contract work was completed in mid-July. The District settled with Fanning for extra compensation for Section 6 in a manner similar to the Heyworth settlement for Sections 4 and 5.

SECTION 7

Section 7, just over one mile in length, included three road and one railroad crossings. However, not all was included in one contract. In a separate contract, discussed under the railroad bridge later in this chapter, was the Chicago & North Western Railroad Mayfair Division Bridge, the Oakton Street Bridge substructure, and the short length of channel excavation from the north side of Oakton to the south side of the railroad crossing. (See Figure 14.) In May 1908, a contract was awarded to Frank F. Healey of Chicago (Healey) for the remaining work in the section; that contract included excavation of the channel, maintenance of ditches and levees for surface drainage, substructures for the Howard and Main Street Bridges, superstructures for all three road bridges, and the construction and maintenance of temporary bridges at Howard and Oakton.

Whereas the schedule called for excavating 54,000 cubic yards by the end of July, Healey had made little progress—only 150 cubic yards had been excavated. When Chief Engineer Wisner interviewed Healey before the contract was awarded, Healey confessed to having no equipment but promised that, if given the contract, he would be able to assemble the necessary equipment and labor within 35 days and then process the work in a timely manner. Wisner expressed doubt, but Healey was emphatic that all arrangements were in order awaiting contract award. By the end of June, after the agreed-to 35 days had passed, no equipment was on site. Wisner encouraged Healey to at least hire teams and scrapers to remove topsoil and excavate drainage ditches, but Healey refused, indicating that work by teams was too expensive and the arrival of his excavating machine was imminent.

The first shipment of equipment parts arrived by mid-July, with one more shipment expected soon thereafter, but Healey confessed its whereabouts were unknown. Wisner threatened forfeiture, but held action as Healey promised that he would start with teams and scrapers. Work started late in July and accounted for the 150 cubic yards mentioned above. The chief engineer estimated that the dragline excavator would not be assembled and ready for production until September 1; under the best of circumstances, Healey would be two months past the contract completion date.

The District was obligated to deposit fill to raise the railroad embankment approaches leading up to the bridge and Healey was expected to perform this work, so the delay put the District in jeopardy with the railroad. If the District didn't place the fill in time to meet the railroad's schedule, the agreement provided that the railroad could bring in their own fill and place it at District expense. Wisner reported his doubts about Healey to the trustees, but he didn't recommend any action be taken. The matter was in limbo for several months. Finally, in December, a supplemental contract with Healey was executed relieving Healey from some of the interim deadlines.

Healey's poor performance continued in 1909. Some of the topsoil was sold to and removed by Calvary Cemetery. Wet conditions in the fall further delayed work. Finally, with Healey's poor performance glaring in comparison to the other contractors, the contract was forfeited; in mid-February 1910, with only 13% of the work completed, he was ordered off the site. Rather than award another contract, in February 1910, two dragline excavators were purchased and District labor completed the work. A new Browning revolving dragline with a 100-foot boom and two-cubic-yard capacity bucket, mounted on four four-wheel trucks, soon arrived. The Schnable & Quinn dragline was moved to the site and fitted for work. In March, District labor began excavation, cleaning ditches and culverts, and installing a lighting plant. By July, despite a flood and pumping to clear the channel of water, completion of the contract work reached 58%. By November, when it was time to fill the channel with water, excavation was not complete. The remainder of the excavation was completed by the District using the two dragline excavators with water in the channel.

A dam, which closed the channel at the south end of Section 7 near Howard Street, was left in place until completion of upstream channel excavation and pump testing at the Wilmette Pumping Station. In late March 1911, the dam was removed following completion of the testing.

The contract the District entered into with the Browning Company for the excavator and related equipment included a buy-back provision: the District had to exercise that provision by September 1910 or pay a penalty. The provision was not exercised and the excavator and equipment continued to be used until near the end of the year, but the penalty was not paid. Browning filed a claim; the District filed a counter-claim citing failure to deliver parts and services. In early 1911, the parties settled their differences. In June, the District loaded the excavator and equipment on a rail car on the Mayfair siding, and Browning accepted the shipment. Browning paid for the excavator and equipment, but only an amount equivalent to the owed penalty.

SECTION 8

Section 8 was 4,730 feet long and was initially excavated by National Brick. The 1907 agreement dealing with the section was discussed earlier. Early in 1909, the drainage ditch was completed, and by the end of the year the channel excavation was about 80% complete. Most of the material excavated by National Brick was placed in spoil piles to be used later for brick making. By early 1910, National Brick had completed their excavation and removed their equipment. However, National Brick was not obligated to perform any road or bridge work and the channel excavation was only a rough cut. To finish Section 8 it was divided into Sections 8A and 8B; contracts for each were advertised and awarded.

The agreement with National Brick required the company to remove 100,000 cubic yards of clay from the channel prism in Section 8, but field measurements of the quantity removed were not made. After completion of excavation by other contractors, the District believed that the full quantity was not removed. National Brick made an offer: they verified the quantity removed, they offered to take an additional 40,000 cubic yards of clay spoil from the right-of-way north of Howard Street, and they requested access to the channel near its plant to build a dock and install drains from their property. The offer was accepted. The method of estimating the originally required quantity of excavation was based on brick production methods; one cubic yard of clay produced 550 bricks. After production records were examined and the calculation made, the District didn't question the amount of excavation.

In May 1913, a permit was issued to National Brick for the extension of the loading and unloading structure across the channel at an elevation 34 feet above the water line. The structure allowed movement of spoil across the channel to the brick production facilities. The working relationship between the District and National Brick proved to be mutually beneficial.

SECTION 8A

Section 8A, which included the Touhy Avenue crossing, made up more than half the Section 8 length. In July 1909, a contract was awarded to Heyworth only for completing the excavation of the channel and surface drainage ditches. Heyworth immediately began excavating the west drainage ditch, working day and night and completing it by the end of the year. In March 1910, Heyworth began excavating the channel working south from Howard Street and alternating between the east and west sides. The channel excavation was completed by early June and the work area cleaned up in July.

SECTION 8B

Section 8B, which had no road crossings, included the remainder of Section 8. In October 1909, a contract was awarded to National Contracting with requirements in the contract similar to those for Section 8A. In April 1910, National Contracting mobilized their equipment and began excavation. Using cableway and dragline, the work progressed quickly and was completed by mid-August. Like Heyworth, National Contracting demobilized and was off the work site the following month.

SECTION 9

The contract for Section 9, a 5,100-foot-long section, was awarded to Heyworth in March 1909; it included the Devon Avenue crossing. In addition to excavation of the channel and construction of ditches and levees for surface drainage, Heyworth was responsible for the construction of concrete abutments and piers for the Devon Avenue Bridge, erection of the bridge superstructure, and the construction of a temporary bridge. Prior to contract award, the District had begun excavating the west drainage ditch; that work continued until mid-year when Heyworth, which had been busy finishing Section 8A, mobilized and was ready to start work. Late in 1909, finishing the work on the west drainage ditch was the first priority; early in 1910, the next priority was clearing and removing stumps in the wooded area north of Lincoln Avenue. Channel excavation with a dragline began on the east side at the south end of the section in March and progressed rapidly, completing 50% by mid-year. Heyworth brought in another excavator and completed the channel excavation by late September. Work on the bridge, explained later, was going on throughout 1910. By the end of the year, Heyworth had demobilized and cleared the site.

SECTIONS 10 AND 11

Sections 10 and 11, which extended over a distance of 4,960 feet, included road crossings for Bryn Mawr, Lincoln, and Peterson Avenues. The trustees were hopeful the City would agree to the rerouting of Lincoln Avenue and elimination of one bridge crossing, so the excavation of the channel and drainage ditch for these sections was not put out for competitive bids but was undertaken by District labor. Work began in October 1908 using steam shovels. Draglines were more efficient at excavation than steam shovels, but the District owned many steam shovels. Later, when the Healey contract was forfeited, the District would purchase the draglines used in that section.

In January 1909, the District began excavating at the south end of the section, but due to persistent cold, wet weather that lasted through March, they lost time. By continuous pumping, the cut was gradually emptied of its water in April, but a late-April storm filled it again. In early May, continuous pumping resumed while track was laid, and excavation with a steam shovel began starting at Bryn Mawr and continuing southward. Most of the spoil was rail-hauled to and deposited in the Illinois Brick clay pit located at Lincoln and Peterson Avenues, with the remainder deposited south of Bryn Mawr and west of the channel. Some spoil was used as fill for the roadway that would become Jersey Avenue and other local road projects. It wasn't until November that rain would cause more delay, but the year ended well with 57% of the excavation completed.

The year 1910 was one of steady progress, with rain delays only in May. To protect the work area from inundation by North Branch floodwater backing up from the south, earthen dams were placed successively at Thorndale and Balmoral Avenues. The District also attached names to their steam shovels: *Chicago*, *Evanston*, and *Ravenswood*. By late November excavation was completed and in December the work site was cleared of equipment. The dam at Balmoral was removed, spoil was hauled to the clay pit by rail, and the removal of equipment began.

SECTION 12 AND THE NORTH BRANCH DAM

Section 12 was the last and southernmost section. It was 4,010 feet long and not only did it connect to the North Branch, but half its length replaced the natural channel. Three road crossings were included, at Argyle Street, and Foster and Lawrence Avenues. The North Branch presented the potential for flooding and required the construction of a concrete dam on the North Branch where it connected to the new channel. Rather than using District labor, Chief Engineer Wisner favored a contract; many contractors had the necessary equipment, they would be interested in bidding, and all of the land was acquired, so he believed it was an opportune time.

In November 1908, seven bids were opened; Wisner's recommendation was to reject them all due to what he explained as irregularities in the bids. The following month five bids were opened and the contract was awarded to Heyworth. In the first bidding, Heyworth came in as fourth lowest bidder; the lowest bidder was a company for which the District had not previously contracted. In the second round the original lowest bidder did not bid; Heyworth cut their unit cost for excavation by 10% but raised other bid costs, which resulted in a second round low bid higher than the first round. While the cost was a little more, Wisner was comfortable with Heyworth, a contractor who was already performing well on other contracts.

Heyworth excavated the channel, constructed a concrete dam for the North Branch and three bridges, and maintained unimpeded flow in the river. All excavated spoil was deposited in designated spoil areas, but Heyworth was allowed to haul spoil off-site at his own expense. Being near developing areas in the city, fill was always in demand. Spoil slopes were no steeper than one vertical on one-and-one-half horizontal, and the toe of the spoil pile was no closer than 35 feet from the top of the channel bank. The channel side slopes were three vertical on five horizontal, but were less steep where ordered by the District. Electric power was provided to Heyworth from the recently constructed transmission line to the Wilmette Pumping Station. Temporary bridges were built at Argyle and Foster for use while the permanent bridges were under construction. Heyworth also constructed the substructure and superstructure for the two bridges with the structural steel supplied by another District contractor. The Lawrence Avenue Bridge was raised on extended piers and approaches completed.

In January 1909, Heyworth began mobilizing for work, but excavation didn't begin until June, the company working day and night northward from Lawrence Avenue. At the river, a diversion channel and a levee were built to separate the river from the construction site of the dam and new channel. By summer 1909, Heyworth was pumping water out of the site for the dam and had excavated half of the new channel south of Foster. Excavation of the new channel in the vicinity of Argyle was slow due to hard clay and equipment breakdowns, but contract work was 51% complete by the end of the year.

By October, the concrete foundation and floor of the dam were completed and work began on the end walls, buttresses, and spillway. Heyworth reasoned that construction of the dam during winter would minimize the flooding potential. The wood forms were stripped off the dam and stiff blue clay backfill was placed around the dam, thoroughly compacted to be watertight. The levee was breached to flood the site of the dam to float out ice and wood. A major setback occurred when the southeast wing wall of the dam failed and the lateral movement damaged a piece of equipment. The levee was restored and the site again pumped out to repair the damaged wing wall. Triple Wakefield piling was driven at the edge of the upstream and downstream sides of the floor to protect against erosion and undercutting the dam. When this work was complete, the levee was breached again in March and water passed over the spillway for the first time.

During wet weather the flow over the spillway was strong, impinging on the east bank of the new channel. It was obvious that, over time, erosion would occur and lead to failure of the east bank. In May 1910, armoring of the east bank opposite the dam was performed by District labor, rather than by Heyworth as a contract extra. Riprap was delivered by scow from Lemont and, aided by a land-based derrick, was placed by manual labor. Completing the excavation and grading the side slopes of the channel and spoil banks occupied Heyworth until summer; all contract work was completed in July. With connection to the North Branch, water in Section 12 backed up to the dam at Balmoral where it awaited the completion of Sections 10 and 11.

Bridges

Although the North Shore Channel was slightly less than eight miles in length, a large number of bridges had to be built: three railroad and 22 road bridges. Railroad bridges will be discussed first, followed by road bridges, in a north to south order. However, the Chicago & North Western Railroad Company Mayfair Division Bridge and the Oakton Street Bridge are discussed together. Because of the proximity of these two bridges, their construction was in one contract. All road bridge design was performed by the engineering department.

RAILROAD BRIDGES

Contrary to the experience on the Sanitary & Ship Canal, the two railroads worked cooperatively with the District. Condemnation was initiated by the District to insure the railroads knew the District was serious in constructing the channel and bridges. To show good faith to the court and the railroads, the District presented proposed agreements with each railroad and the trustees adopted resolutions expressing intent to execute the agreement. The District was allowed to cross the right-of-way of each railroad with a channel and construct the channel—and was required to reimburse the railroads for their costs in construction of permanent and temporary bridges.

CHICAGO, MILWAUKEE & ST. PAUL RAILROAD BRIDGE

In 1907, the right-of-way of the Chicago, Milwaukee & St. Paul Railroad, presently the Chicago Transit Authority Purple Line, was leased to the Chicago & Milwaukee Electric Railroad. The tracks had not yet been elevated so trains ran on grade. The District constructed a temporary diversion for the tracks around the permanent bridge construction site, designed and constructed the permanent bridge, and compensated the railroad for ongoing maintenance and repair of the bridge. The District also provided a warranty should canal operation cause erosion and settlement of the bridge foundation. The railroad approved the design of the bridge, taking ownership of the bridge upon its completion and acceptance by the railroad. The original bridge remains in service.

By September 1908, the track diversion was completed, allowing channel excavation to proceed in the vicinity of the bridge site. In December, a slope failure of the diversion embankment occurred and pilings were driven to support the slope to prevent further slope failures. Also in December, a contract for design and construction of the bridge superstructure was awarded to the Wisconsin Bridge & Iron Company of Milwaukee for a double-track, riveted, 175-foot-long through truss spanning the channel, along with two 55-foot-long deck plate girder approach spans at each end of the truss. The superstructure contract was awarded before the substructure contract; that was helpful to the engineering department in designing the proper foundations for the bridge. In April 1909, a contract was awarded to National Contracting for furnishing and driving concrete piles and for forming and pouring concrete pile caps and abutments for the bridge substructure.

In June, driving piles for the north abutment began and work followed on the north pier, south pier, and south abutment; substructure work was finished in August. Bridge superstructure work, already begun on the north approach span, was followed by work on the south approach span, which in turn was followed by work on the through truss. The Chicago & Milwaukee Electric Railroad began train service on the new bridge in September, and in October 1909 all superstructure work was completed.

CHICAGO & NORTH WESTERN RAILROAD BRIDGES

In March 1908, an agreement with the Chicago & North Western Railroad was executed to provide for bridges at two locations where the channel crossed railroad right-of-way: the Milwaukee Division of the railroad near West Railroad Avenue and the Mayfair Division, sometimes called the Mayfair branch or Mayfair cutoff, near Oakton Street. A four-track bridge was required for the Milwaukee Division and a three-track bridge was required for the Mayfair Division. The terms and conditions of the agreement were similar to the agreement with the Chicago, Milwaukee & St. Paul Railroad.

CHICAGO & NORTH WESTERN RAILROAD MILWAUKEE DIVISION BRIDGE

The design of a stone arch bridge, though well along, was abandoned in favor of a steel plate girder bridge with concrete abutments and piers. A stone arch would have been too heavy for the underlying soils and foundation piers. The steel plate girder bridge was built, and it remains in use to the present time. The bridge's name plate is still on the side of the bridge girder facing Green Bay Road, but it is not safe to climb to a position to observe it.

In December 1908, the Evanston city council adopted an ordinance providing for the construction of the bridges for the railroad and West Railroad Avenue. The District constructed two new bridges, two temporary bridge trestles, and two diversions, and restored the railroad and road area upon completion. The diversion for the railroad was to the west in the right-of-way of West Railroad Avenue; the road diversion was west of the road right-of-way. All construction was at District expense.

In January 1909, design and construction of the railroad deck plate girder superstructure spans were awarded to George W. Jackson, Inc. of Chicago (George Jackson). During design, District work on the substructure was encountering difficulty due to the soft unstable clay subsoil. In August, the railroad requested that they be allowed to complete the construction of the substructure at stated prices. The chief engineer, judging the prices reasonable, was of the opinion that it would be in the District's best interest to work with the railroad. An agreement was executed modifying the original March 1908 plan and in November the chief engineer was authorized to allow the railroad to proceed.

In April 1909, after the District constructed the fill for the track diversion and the temporary trestle, the railroad laid their tracks around the site for the new bridge. In August, the District began substructure foundation excavation, but was delayed by rain and flooding. Before excavation for the bridge piers was begun, a row of protection piles was driven north of the north abutment to prevent a slope failure. However, a day after excavation began a slope failure occurred that filled in the pier excavation. As a result, excavation was begun for the south abutment and south pier while the slope failure was removed and the north back slope was made less steep. As a precaution the south slope was also cut back. Work continued on the abutments, but excavation was delayed by rain and ground frost.

The railroad took over the substructure work; by the end of the year about 70% of the substructure was completed. George Jackson was assembling and riveting the bridge span girders to the side of the site, pending abutment completion. In January 1910, the railroad worked on the north and south abutments, excavating, building forms, and pouring concrete. In February, both abutments were completed, allowing Jackson to begin and complete installation of the girders. In April, following installation of tracks, the rails and ties were removed from the diversion. Under the channel bottom, between bridge piers and abutments, the District excavated, built forms, and placed concrete for struts. The railroad backfilled the abutments, graded the track bed, and laid rails; by October 1910, all work was completed.

CHICAGO & NORTH WESTERN RAILROAD MAYFAIR DIVISION BRIDGE AND OAKTON STREET BRIDGE

It was a busy construction site, made complex by the proximity of the two bridges, the intersection of the railroad and street east of the channel and the location of a rail yard east of the channel, and north of Oakton Street. The Weber rail yard was used as a freight terminal and for storage of coaches used for commuter trains on the Milwaukee Division Line. The agreement between the District and the railroad laid out work to be performed by each party. The two bridges and channel excavation between the two bridges, which were in the midst of Section 7, were combined into one construction contract, The north and south approaches to the railroad bridge were raised at 0.3% grade for 2,500 feet on each side of the bridge, so the railroad tracks could be elevated 7.5 feet at the bridge to meet the clearance above the water level. The railroad performed all work in diverting the tracks to the east around the railroad bridge construction site: they raised the grade of the Division Line and customer sidings, raised the grade of the Oakton Street railroad crossing, and installed the track on the new bridge and approaches. (See Figure 14.)

In September 1908, the railroad began work on the track diversion; in the following month, the earth work was completed and the tracks were moved from the mainline to the diversion. Over the next several months, the railroad performed track work in the rail yard; at the same time, the bridge substructure and superstructure was being put under contract and then constructed. In October 1908, a contract was awarded to Fanning for the substructures of the two bridges and intermediate excavation. Because of the close proximity of the two bridges and the need to keep the railroad running, the contract defined the sequence of construction. Fanning excavated for the bridge abutments and piers, built forms and poured concrete for the abutments and piers, and excavated about 1,000 feet of channel. They followed the prescribed sequence of work: excavation of the channel at the railroad bridge site; construction of the piers for the railroad bridge abutments; rerouting of drains, utilities, and roadway at the Oakton Street crossing; construction of piers for the Oakton Street bridge foundation; construction of railroad bridge abutments; and construction of Oakton Street bridge abutments.

Early in 1909, Fanning began excavating the foundation piers for both bridges; they used a hoisting engine and derrick, with manual labor below ground excavating the cylindrical shaft to bedrock. Each bridge had piers on each side of the channel. As shaft excavation proceeded, metal shields held the native soil in place until the shaft was filled with concrete. Concrete abutments were formed for the railroad bridge and poured on top of the piers. For Oakton Street the piers extended higher, with a concrete pier cap to support the superstructure; abutments to support the bridge approach spans were built on native soil at the top of the channel bank. In addition, to resist lateral soil pressure, concrete struts were built beneath the channel bottom between the railroad bridge abutments. The channel was excavated from the north side of Oakton Street to the south side of the railroad bridge. Fanning completed the all substructure work in November.

Per specifications, the District ordered Fanning to extend all piers to bedrock rather than to hardpan clay. By bearing on bedrock, the potential for future settling of the foundation would be averted. To lessen the risk of slope failure, Fanning was also ordered to perform additional excavation to flatten the side slopes of the channel near the bridge abutments.

In January 1909, George Jackson was awarded the contract for the superstructure of the railroad bridge. For the three tracks, Jackson furnished and erected the steel plate girder deck bridge with ballasted floor and a low-steel clearance of 16 feet above CCD. The steel girders were seated on the abutment built under the Fanning contract. In September, Jackson began work erecting the superstructure and, except for the wreck of a derrick, made steady progress. By November the floor of the new bridge was sealed with asphalt and the project was finished.

In October, the railroad resumed work, finishing the raising of the north and south grades approaching the bridge. In the following month, by mid-November, ballast was placed on the bridge floor, the north bound track was laid, and the first train crossed the new bridge. The south bound track was laid four days later and, on the following day, the first south bound train passed. Track and ties on the diversion were taken up and used elsewhere; signal lines were relocated.

The Northwestern Gas Light & Coke Company (Northwestern Gas) operated a manufactured gas plant on the south side of Oakton Street west of the channel. For service in Evanston, they wished to maintain their gas main running eastward on Oakton. The District offered to allow the pipeline to be suspended from the bridge, but Northwestern Gas declined for safety reasons; they preferred the gas pipeline to cross the channel in a tunnel 35 feet below CCD. In March 1910,

Northwestern Gas diverted a gas main at Oakton Street for channel construction; later in June they started building the tunnel beneath the channel along the south side of Oakton. In the same month, the railroad replaced the switch for the Budlong siding and added ballast along the track. The Budlong siding referred to a spur track, known as the Chicago & West Ridge Railroad, which ran south along the east side of the channel from Oakton Street to Lincoln Avenue to service agricultural and industrial customers. In August, the District excavated, built forms, and placed concrete for struts between the piers of the Oakton Street Bridge. Roemheld, the superstructure contractor, hauled steel from the Lincoln Avenue rail siding and began and completed steel erection. The District graded the approaches, and in late September the bridge was completed.

The Mayfair Division was originally built in 1880s to bypass freight from the Milwaukee Division, which was a main line for passenger service into downtown Chicago. In 1958, the Weber rail yard on the north side of Oakton Street was closed, but freight trains travelled the line into the 1960s. In 1989 the railroad decommissioned the Mayfair Division and removed the tracks and signals; the bridge girder spans remain in place.

ROAD BRIDGES

With the exception of the Sheridan Road Bridge, all road bridges over the channel were of a similar design. While all other road bridges were free standing, the Sheridan Road Bridge was integrated with the pumping station and navigation lock. Typically, the bridges had three spans, two approach spans over each channel bank or side slope, and one center span over the channel water surface. The far ends of the approach spans were seated on abutments at the top of each channel bank; the connections of the approaches and channel span were seated on transverse caps attached to the top of the vertical piers located on each side of the channel water surface. As an alternate design, the transverse cap was nearer to the water line and a structural steel bent in truss form was used to support the connections of approach and channel spans.

The bottoms of the piers were founded on hardpan clay or bedrock. The width of the bridge varied based on the traffic load—either two or four lanes, with sidewalks on each outer side. The crossing was skewed in a few cases, such as Lincoln Avenue in Chicago and Lincoln Street in Evanston, but the design remained basically the same with the piers offset to align with the channel direction. Roadways were either 24 or 36 feet wide, and sidewalks were either five or six feet wide. Few of the original road bridges remain; over the years the more heavily travelled bridges have been replaced with larger structures. The only bridges remaining the same as built are at Linden and Maple Avenues in Wilmette.

The bridges were built in different ways, but generally the substructures and superstructures were in separate contracts. Some substructures were constructed by the section contractor or the District, some by a separate contract. Separate contracts were issued for fabrication and delivery of the structural steel and for erection and completion of the bridge. Further complicating the matter, fabrication and erection were in groups or in separate contracts. (See Table 2.) Following is a chronological list of contracts for furnishing, fabricating, and delivering superstructure metalwork:

- June 1908: Indiana Bridge Company of Muncie, Indiana, for one bridge
- June 1908: Toledo-Massillon Bridge Company of Toledo, Ohio, for 11 bridges
- June 1908: Wisconsin Bridge & Iron Company of Milwaukee, Wisconsin, for one bridge
- May 1909: W.A. Jackson of Chicago, for five bridges
- July 1909: Fort Pitt Bridge Works of Pittsburgh, Pennsylvania, for one bridge
- July 1909: Kenwood Bridge Company of Chicago, for one bridge

The superstructures for eight bridges were erected and completed by the section contractor. Contracts limited to superstructure erection and completion were awarded as follows:

- October 1908: Kelly-Atkinson Construction Company of Chicago, for four bridges
- September 1909: Charles Volkmann & Company of Chicago, for two bridges

- September 1909: Roemheld Construction Company of Chicago, for six bridges

The metalwork was supplied to the erection contractors; the contractors furnished all other materials necessary for completion of the bridge. For Lawrence Avenue, the existing bridge span was raised on the extended existing piers; the bridge was completed by Heyworth, the section contractor.

LINDEN AVENUE BRIDGE

In September 1908, the District began excavation for the Linden Avenue Bridge abutments and pier shafts; construction of the substructure was completed in November. In December, the superstructure steel was delivered and Kelly-Atkinson began erection. All steel was erected and by January 1909 the bridge was completed; in February the bridge was opened for traffic; and in August 1910, the District excavated the channel bottom, built forms, and placed concrete for the struts between the bridge piers.

HILL STREET BRIDGE (RENAMED MAPLE AVENUE BRIDGE)

In December 1908, all foundation work was completed by District labor. In the same month, Kelly-Atkinson began steel erection, which was completed in January 1909. The bridge opened to traffic in February. For the struts between the piers, the District began excavation and placing concrete in October, and in November the work was completed.

ISABELLA STREET BRIDGE

In July 1909, District labor excavated the foundation pier shafts and poured concrete. The belled bottom of the piers extended into hard pan clay at 46.3 feet below CCD. In September, the east and west abutments and piers for the bridge were completed. In November, Volkman began superstructure steel erection, which was nearly complete the following month. In January 1910, bridge deck flooring was completed; the bridge opened to foot traffic on January 10. In November, the District excavated the channel bottom and placed concrete for the struts between the bridge piers, completing the work by late November.

CENTRAL STREET BRIDGE

In accord with a September 1908 agreement between the District and the Chicago Consolidated Transit Company, the District used excavated spoil from the channel to create a road diversion approximately 150 feet south of Central Street; trolleys began running on the diversion by October 1908. The District reimbursed the company for its costs for track relocation. By the end of 1908, the bridge abutments had been completed, and in January 1909 a derrick and hoist were erected for excavation of the bridge pier shafts. In early March, the west shafts were completed, using worn rails for reinforcement of the concrete piers. The superstructure erection was completed by Kelly-Atkinson, and by late March the traction company began laying rails on the bridge floor. In April 1909 the District completed paving the approaches, and the bridge was completed and opened to traffic in May. In October 1910, the District began work on the struts between the piers; all work was completed by late November.

LINCOLN STREET BRIDGE (EVANSTON)

In September 1909, the District began excavation for the east and west abutments, and in October the bridge pier shaft excavation began. No work occurred on the bridge over the winter and early spring. East abutment concrete work was completed in June 1910; west abutment work completed the following month; and the piers were completed in October. Also in October, Volkmann began and completed steel erection, then the District began work on the struts between the piers; all work was completed by late November.

WEST RAILROAD AVENUE BRIDGE (RENAMED GREEN BAY ROAD BRIDGE)

Construction of the permanent bridge followed completion of the Chicago & North Western Railroad bridge and removal of the track diversion; that construction was immediately to the northeast. In March 1910, Schnable & Quinn began excavating for the pier shafts. Meanwhile, following completion of the railroad bridge, the telephone company was restoring lines and building two vaults for lines to cross the channel. In May, substructure work was stopped pending design changes. Until this time the road had been closed with traffic detoured via Dewey Avenue. In June, immediately to the west of the road, Schnable & Quinn built a diversion embankment and trestle and the use of Dewey Avenue ceased. Substructure work resumed and was completed in July.

Schnable & Quinn began superstructure erection, completing the framing in August, and the District completed roadway approach grading. The bridge wing walls and the struts between piers were not included in the Schnable & Quinn contract, so the District began excavation for those tasks. Evanston completed the suspension of a water main from the bridge, allowing Schnable & Quinn to complete the bridge. Approach roadway paving, the concrete wing walls and the sidewalk between Grant and Noyes Streets were completed; by mid-November the bridge was opened to traffic.

BROWN AVENUE BRIDGE (RENAMED BRIDGE AVENUE BRIDGE)

After completing the bridge substructure piers and abutments, in July 1909, Heyworth moved its superstructure erection equipment to the bridge site and began erection. Steel erection and painting were completed a month later as deck flooring was being installed. The superstructure was completed by September. The approaches and roadway leading from Simpson Street on the south and Noyes Street on the north were graded and the bridge opened to traffic. In October, to dewater the area under the bridge for excavating, Heyworth built dams in the channel on each side of the bridge so they could build forms and pour concrete for the struts between bridge piers.

EMERSON STREET BRIDGE

By September 1908, Heyworth completed the substructure work and began superstructure steel erection. Work, stopped over the long winter season, resumed in April. Heyworth completed steel erection, and in May 1909 was beginning to install the decking. However, due to lumber shortages the flooring couldn't be completed. Supplies resumed in July and the flooring and painting was completed. Work on the east approach was held pending the moving of a dragline excavator, after which grading was completed. In October, to dewater the area under the bridge, Heyworth built dams on each side of the bridge so they could build the struts between bridge piers.

CHURCH STREET BRIDGE

In September 1908, the Church Street substructure work was well underway and Heyworth included the pier struts so it wouldn't be necessary to come back later, as had happened at other bridges. Since all steel delivery was completed, Heyworth began steel erection. As steel erection was nearing completion, Heyworth's riveting compressor plant was destroyed by fire. Fortunately, the oak flooring had not yet been delivered so it was not consumed in the blaze. Work resumed, and in January 1909 the framing, painting, and flooring was completed. The Church Street Bridge opened in early February.

DEMPSTER STREET BRIDGE

In October 1908, as one of the first tasks in his contract for Section 5, Heyworth began work on the abutment and pier foundations for the bridge. Steel erection began and, by the end of October 1909, was nearly complete. In the following month, Heyworth, working on the struts between piers, had problems with excess water in the channel bottom so dams were built to isolate the work area. In December, Heyworth continued working on the pier struts, but continued to have delays due to wet conditions. All work on the bridge was completed in February 1910.

MAIN STREET BRIDGE

In September 1909, Healey began excavation and pouring concrete for the substructure piers, completing them the following month. Because of cold wet weather and contract forfeiture, Healey didn't complete the abutments; District labor completed Healey's abutment excavation in March 1910 and completed the abutments in April. In September, with the struts between piers completed by the District, Roemheld hauled the superstructure steel from the Lincoln Avenue rail siding. In October, District labor completed the work on the piers that Healey had begun, then Roemheld began steel erection. The bridge was completed and opened to traffic at the end of November.

HOWARD STREET BRIDGE

In October 1909, Healey began excavation and concrete pours for the piers and abutments, but didn't complete this work. In February 1910, following the forfeiture of Healey's contract, the District took over and by May had completed the substructure. In June, superstructure erector Roemheld began work, and in July the framing was nearly complete. In August the District excavated and placed concrete for the struts between piers and graded the bridge approaches. Roemheld completed bridge erection except for some faulty rivet placement; rivet replacement was completed in September.

TOUHY AVENUE BRIDGE

In May 1909, District labor started substructure construction and completed it the following month. Steel erection had not been included in a group contract or the Heyworth contract, so in April a separate contract was awarded to Kelly-Atkinson. In July, steel erection began, and was completed in August. Heyworth isolated the channel under the bridge with two dams, then dewatered the area, excavated for the struts between piers, built forms, and placed concrete. The struts were completed by the end of September.

DEVON AVENUE BRIDGE

In June 1910, Heyworth began substructure construction and completed all substructure and superstructure work in time for the bridge opening in mid-August. Beneath the bridge, Heyworth completed the struts between the piers in September.

LINCOLN AVENUE BRIDGE (CHICAGO)

District labor completed the substructure work in October 1910, and in November Roemheld began superstructure erection. With Roemheld working overhead, District labor constructed the struts between the bridge piers; the bridge opened in December.

PETERSON AVENUE BRIDGE

In June 1910, the east abutment was completed and in July all substructure work was completed. Superstructure work commenced and at the end of August the bridge was completed. District labor then backfilled and graded the approaches; the bridge opened in mid-September.

BRYN MAWR AVENUE BRIDGE

In August 1910, District labor began work on the substructure and in October that work and the struts were completed. Roemheld began steel erection in the same month, completing the bridge in November. It opened to traffic in December.

FOSTER AVENUE BRIDGE

In February 1910, Heyworth, while excavating the pier shafts adjacent to the existing bridge, encountered water-bearing sand and gravel. That stopped the work until steel cylindrical shields could be obtained for the excavated shafts. The shafts were excavated to hardpan clay; the bases were belled to twice the diameter of the pier and the shaft was filled with concrete. In March, Heyworth drove piles and erected a temporary bridge, which opened by the end of the month. Reinforced concrete permanent bridge abutments, both east and west, were built in native soil. Meanwhile, utility work required coordinated efforts to clear the site: the Chicago Telephone Company relocated their communication wires; Peoples Gas, Light and Coke Company removed an old gas main; and the City installed a water main suspended from the temporary bridge. In May, Heyworth completed the abutments, piers, and struts. After the structural steel was delivered by W.A. Jackson, Heyworth started in June to erect the superstructure, completing the bridge in July. Heyworth graded the approaches, and in late July the bridge opened to traffic.

In September 1911, upon a request by the City, the District paved the sidewalks on the bridge and approaches via a contract awarded to Citizens' Construction Company of Chicago.

ARGYLE STREET BRIDGE

In June 1909, the existing bridge was dismantled and in July Heyworth completed all substructure piers. Due to Heyworth's priorities on other tasks for the Section 12 contract, work was drawn out. In April of the following year, the abutments were completed and steel erection began for the superstructure. In May, the roadway approaches were graded and paved, but work was stopped by an iron workers strike. Erection resumed by July, and the bridge was completed and opened to traffic in August.

LAWRENCE AVENUE BRIDGE

Heyworth raised the existing steel truss bridge and extended the foundation piers to provide for a low steel clearance of 16 feet above CCD. By November 1909, Heyworth completed all bridge details, including raising the grade of the roadway approaches. After years of standing incomplete, the bridge opened; Lawrence Avenue was finally continuous over the channel.

Road Bridge Maintenance

Traffic on the bridges had been more than anticipated, which caused excessive wear of the wooden flooring. By 1913, repairs were made, replacing pine or spruce with oak flooring or, for some bridges, with steel grating. Beginning in 1914, electric lighting was installed upon request of the local municipality. The east abutment of the Hill Street Bridge had moved toward the channel, damaging the piers and a sewer outfall in August 1916; repairs were undertaken immediately. In subsequent years, bridge maintenance became an ongoing activity for the District, particularly the periodic need to replace the oak flooring. Except for six bridges, ownership of each bridge was transferred to the local municipality. As area development occurred and traffic increased, bridge owners replaced entire bridges with larger structures.

In present times, the District continues to own and maintain six bridges: Church and Main Streets and Touhy Avenue in Skokie, and Linden and Maple Avenues and Sheridan Road in Wilmette. The three bridges in Skokie have been rebuilt and modernized. The Illinois Department of Transportation currently requires periodic structural inspections and reporting by the bridge owners. The District has proactively maintained the structural integrity of these six bridges.

Channel Side Slopes

Side slope failures due to the soft clay subsoil were ongoing, and various methods were attempted to address the problem. A contract was advertised in 1912 for the installation of 1,600 lineal feet of channel side slope paving with rock riprap south of Peterson Avenue, including slope paving with concrete under three bridges, but all bids were rejected as too costly. Instead, Great Lakes was awarded a contract for dredging critical parts of the channel; the spoil to be deposited either east of the channel at Lincoln and Peterson Avenues in the partially filled Illinois Brick clay pit or dumped from scows into Wilmette Harbor with the intent to move it to the landfill adjacent to the harbor. However, before Great Lakes could move the spoil to the landfill the Wilmette Park District, now the owner of the landfill, requested that no more material be placed in the area behind the crib wall. The spoil in the harbor was removed and, when weather allowed, disposed in designated areas in the lake.

Meanwhile, to test various methods, the District performed the slope paving using its own labor and equipment on small demonstration projects. In 1913, the area of side slope paving near Peterson Avenue was enlarged because the initial installation of paving with riprap, combined with piling driven at the water line, appeared to improve stability of the channel bank. In October, plans were prepared for a 500-foot demonstration project on each side of the channel from Sheridan Road to Linden Avenue; that project would have side slope protection by riprap paving above the water line, held in place by a double row of piles at the base of the paving. A contract for that work was awarded in November to Great Lakes. A mild winter allowed the work to continue into 1914, and it was completed.

Frustration with frequent slope stability failures on the North Shore Channel, and similar experience on recently awarded contracts for construction of the Calumet-Sag Channel, led the trustees to form a commission to study the problem, inspect similar construction problems in other areas, and report findings and recommendations. A brief summary of the commission's work is found in Appendix A-6. Along the channel, unstable slopes and erosion also resulted in separation of sewer outfall structures from the sewer, requiring frequent repairs by District labor.

Early in 1914, it was estimated that 96,000 cubic yards had sloughed off the banks into the channel, with an additional 50,000 cubic yards added each year from sloughing induced by the winter freeze-thaw cycle. The situation was dire; the trustees needed a solution or the channel could be declared a failure. Steeper slopes invited failure and numerous failures had occurred in the reach between West Railroad and Linden Avenues; it was concluded that making the side slopes less steep would result in a reduction in the frequency of failures. The earlier demonstration project had proved successful, so a project at the water line lake ward of Emerson Street was planned to flatten the side slopes and install more timber pile protection.

In July 1915, a contract was awarded to Byrne Brothers Dredging and Engineering Company of Chicago to remove all spoil remaining on the top of the channel bank, excavate soil on the bank to achieve the new flatter slope of one vertical on three horizontal, and install pile and timber protection at the waterline. Spoil removal and flattening the side slope occurred on both channel banks between West Railroad Avenue and Sheridan Road. The timber pile protection was improved to include a row of vertical and batter (slanted) piles, connected at their tops and surmounted by a timber cap and whaling, with a row of sheeting behind the cap. Byrne proceeded, but was ordered to stop work by winter; cold weather would increase the likelihood of slope failure and the District didn't want to take the risk. In addition, the harbor was overloaded with spoil that couldn't be moved to the disposal areas in the lake. Another problem came to light: there had been no permit application for lake spoil disposal.

In 1916, work resumed with a contract awarded to Great Lakes that extended into 1917. The slopes were made less steep using hydraulic sluicing and dredging to remove the sluiced soil from the channel. Channel water, pumped through a hose and a high pressure nozzle, was directed at the face of the soil to be removed. The water jet eroded the soil into the channel. An improved timber pile protection was installed along reaches of the channel slope designated by the District. The combination of flattening the slope and timber piles, and sheeting at the water line, abated side slope failures.

The May 1907 agreement with National Brick required them to move scows to maintain their schedule for spoil pile removal between West Railroad Avenue and Pratt Boulevard, but that was prevented by deposition of sediment in the channel from the side slope failures. In November 1916, a supplemental agreement was executed addressing several issues: the District was obligated to maintain a channel at least 30 feet wide and eight feet deep, National Brick was given a five-year extension to January 1, 1934, the south limit was extended from Pratt Boulevard to Lincoln Avenue, and National Brick could construct up to eight lie-bys, splitting the cost with the District. The lie-bys would facilitate the movement of the scows in the narrower channel.

In 1917, National Brick dredged the channel from West Railroad Avenue to Bridge Street to a depth of nine feet and a bottom width of at least 40 feet. Due to extremely severe winter weather, National Brick also requested another time extension in March but no action was taken.

North Branch Dam and Upstream

To continue the purpose of the dam, several modifications had been made. Following 1910, the modest impoundment behind the dam caused the deposition of sewage solids; as a result, low flow in warm weather lead to obnoxious odors. In the overflow wall, a large rectangular hole was cut lower than the low flow notch in the top of the wall. But this hole became frequently plugged with tree limbs and debris, which required more clearing to prevent odors. In the late 1920s, by cutting out the concrete above the hole, the hole was modified to become a deep notch. Less frequent cleaning was necessary, except when large tree limbs spanned the notch to catch debris and cause odors. In the late 1960s, the notch was widened; that appeared to resolve the problem, but occasional cleaning by the District is still necessary.

In the late 1930s, a federal Works Progress Administration project implemented by the City paved the stream bottom for about a half mile upstream of the dam that eliminated the soft sediments on the river bed and increased discharge capacity. In many locations, however, over the course of time, the pavement had broken up or become overlain with sediment, defeating the intended purpose of increasing capacity. Throughout the river course in the forest preserves, tree fall in the river was thought to be a reason for diminished discharge capacity and increased flooding. Citizen activists, like Ralph Frese with the Cook County Clean Streams Committee, worked to remove debris, but volunteer efforts weren't enough. In the 1970s, urging by the local representative in Congress resulted in a multi-year project by the Corps to clear the river channel. Stream maintenance is currently performed cooperatively by the Forest Preserve District and the District.

For several years the Friends have advocated for replacement of the North Branch Dam to allow upstream fish migration. The likelihood of fish migration has scientific support and its occurrence would improve upstream ecology. Fish migration in the downstream direction is not restricted. Whatever structure is built, perhaps a whitewater chute, it will have to be safe over all ranges in flow. The District commissioned the University of Illinois to study various types of structures. Safety was demonstrated to be a concern at high flow, especially in a situation with high flow on the tributary and low flow in the downstream canal. The District built the dam as an adjunct to the North Shore Channel and will have to acquiesce if a decision is reached to modify the structure.

Wilmette Pumping Station and Lock Updated

Between the channel and lake, the lock was rarely used for boat passage other than construction vessels. The District discouraged boat traffic in the channel due to concerns with side slope stability and wave action. The original miter gates pointed in the downstream direction because pumping resulted in the canal water level being above the lake water level. In 1930, even before the previously-described change in canal system operation, the upstream miter gates were not holding securely against the variable lake water level. Rather than modify both sets of miter gates, a slide gate recessed into a pocket in the south wall replaced the east miter gates. The slide gate was more suitable in holding against variable lake levels.

An October 1954 flood was the first event since 1939 when floodwater was intentionally released to the lake through the Chicago River Controlling Works. The miter gates at Wilmette were not designed to release floodwater to the lake. After the July 1957 flood event, when floodwater was intentionally released to the lake for the second time, plans were made to modify the west miter gates.

In August 1959, a contract was awarded to M.J. Boyle & Company for a new lock gate. The new 30-foot-wide, electrically operated roller lift gate was installed to replace the west miter gates; the east slide gate was made inoperable and was secured in its pocket. This improvement was necessary to have a strong gate to hold floodwater in the channel and, when the need arose, a reliable and fast means to open the gate to discharge floodwater. Urbanization in the north suburbs and North Side of the city following World War II increased the frequency and magnitude of floods, and the pumping station could be flooded if the floodwater was not released rapidly. Replacement of the gate was timely; it was first used for floodwater discharge on two occasions in September 1961.

The pump room floor is about 11.5 feet below CCD, and to keep the pump motors and the floor dry, two sump pumps have always been in service to eject seepage. Over a very cold winter holiday weekend in the mid-1960s, the outside discharge from the sump pump froze, causing both sump pump motors to burn out. By the time personnel arrived on the next work day, the pump motors were under water. After extensive repairs were completed a simple rule was put into effect: Wrap the outside discharge pipe with electrical heat tape.

By the arrival of the twenty-first century, the big sluice gate had been used many times; age and wear were taking a toll. The gate had a nasty habit of becoming jammed and unable to open fully. The pumps, though maintained, were approaching a century of service and their performance was not reliable; their age dictated rehabilitation or replacement. Portable pumps were brought in when lake water needed to be pumped into the channel. There was rarely a need for the four large pumps, but there was a need for a better way to control how to bring lake water into the channel or release floodwater to the lake.

In 2012 and 2013, following a contract awarded to F.H. Paschen, S.N. Nielsen, major improvements were made at the pumping station. Two of the four pumps were rehabilitated, variable speed motors were installed, and sluice gates were built to protect the pumps when not in operation. The other two pumps were removed; in their place, gates were installed in each pump tunnel allowing these two tunnels to be used in either direction for gravity-driven passage of water. Screens were installed on all four tunnel inlets for control of debris. The old 30-foot sluice gate was replaced with three smaller hydraulic-lift sluice gates. With smaller gates, the flows could be better modulated and avoid the near tidal wave conditions in the harbor when the single old gate was opened. It is believed that the pumping station is good for another century of service.

Wilmette Harbor Management

The harbor and landfill on the lake bed, which is owned by the public in the name of the state of Illinois, was created by the District. In 1911, the state of Illinois transferred title of the created landfill to the Wilmette Park District for public recreation. A few small parcels abutting the harbor walls and the walls are owned by the District. In 1929, the Wilmette Park District conveyed title of a parcel to the U.S. Coast Guard for establishment of a station for search and rescue operations. In the same year, the Sheridan Shore Yacht Club, a social and sport group that grew out of the boat owners using the harbor, was formally organized. Those who were more focused on the harbor infrastructure, a group that in 1938 became the Wilmette Harbor Association (Harbor Association), took care of maintenance and operations as a service to the boat owners. Both the Harbor Association and yacht club are not-for-profit organizations and lease their land-based premises from the District.

The Harbor Association administers rules for harbor users; it also collects mooring fees and uses the proceeds for maintenance and operations. The association existed without any formal agreement or lease with the District until July 1962, when the new sluice gate was installed in the lock chamber. A lease became necessary to protect the District's liabilities—for use of District property, and for boats moored in the harbor (if they were damaged as a result of District operations). In exchange for a no-fee lease, the Harbor Association assumed the liability and the obligation for maintenance, including periodic dredging. When the District opens the big sluice gate to release floodwater, a virtual tidal wave tears through the harbor and moored boats can be tossed about like toothpicks. To avert such a happening, the Harbor Association clears and secures boats whenever rain is forecast.

In 1964, record low water levels for Lake Michigan occurred, and to assure sustainable depths for water flow and boat mooring, the Harbor Association took the opportunity to remove clay from the harbor bottom. The usual dredging methods were not used; rather, with District consent, the harbor inlet was blocked with an earthen dam and the harbor was drained by pumping. Then the clay was scooped out using draglines and trucked away for disposal outside the harbor.

Another improvement was an extension, farther into the lake, of the inlet channel walls. Five large barges were sunk and filled with rock, three on the north side and two on the south side of the inlet channel; the purpose was to subdue wave action and reduce the accumulation of sand in the inlet channel and harbor. Over the years, harbor walls have been reinforced and sidewalks installed for convenience and safety of boaters and public.

The arrangement was mutually beneficial, and in 2012 the 50-year lease needed to be renewed. In the intervening years, the District statute was changed to require competitive bidding for non-municipal leases. In 2009, the Harbor Association requested a meeting with the District to find out about renewal under the new requirements. While the District normally begins a lease renewal one year prior to termination, one year is not enough time to deal with annual dredging, liability insurance, and 250 boat owners. The Harbor Association, accommodated by the District, began a tenuous process to find a way to maintain this mutually beneficial arrangement and comply with the statute. Appraisals were obtained and the new lease was advertised.

Leases for District land are usually noncontroversial as District land is prized by few, so it was a surprise when three bids were submitted. Eventually, in April 2013, a new lease was awarded to the Harbor Association, the only bidder of the three determined to be qualified. Under the new lease, the District is in even better shape—the Harbor Association pays a significant annual rental fee determined through appraisals and the competitive process, and they are still responsible for liability and maintenance.

The North Shore Channel, an attractive and popular part of the canal system, is tree-lined throughout most of its length and it maintains a tranquil beauty for canoes, kayaks, and small recreational boats. Lacking disturbances caused by large motorized boats, the canal banks are not destructively scoured by waves; it has the best potential for aquatic habitat, a potential that is generally lacking throughout the remainder of the canal system.

References

Chicago Tribune Archives. archives.chicagotribune.com.

Cleveland, Jessica. *The North Shore Channel: From Vision to Reality*. Unpublished manuscript. Evanston, Illinois, 2007.

Ebner, Michael H. *Creating Chicago's North Shore*. Chicago: University of Chicago Press, 1988.

Evanston History Center. Railroad Transportation files.

Hoppe, Heidrun, and Jack Weiss. *Evanston: 150 Years, 150 Places*. Evanston, IL: Design Evanston, 2013.

Illinois Secretary of State. cyberdriveillinois.com.

Metropolitan Sanitary District of Greater Chicago. *The Story of the Metropolitan Sanitary District of Greater Chicago: The Seventh Wonder of American Engineering*, 1959 Edition.

Metropolitan Sanitary District of Greater Chicago. Maintenance and Operations Department Annual Reports, 1954, 1957, and 1961.

Metropolitan Sanitary District of Greater Chicago. Proceedings of the Board of Commissioners/Trustees, 1956 through 1988.

Metropolitan Water Reclamation District of Greater Chicago. Maintenance and Operations Department Facilities Handbook, 2012.

Metropolitan Water Reclamation District of Greater Chicago. Proceedings of the Board of Commissioners, 1989 through 2015.

Piersen, Joe. *C&NW Lines North of Mayfair: Maps*. Deerfield, IL: Chicago & North Western Historical Society, 2004.

Rogers Park/West Ridge Historical Society. *The Historian*. (Summer 2014): 12–13.

Sanitary District of Chicago. Proceedings of the Board of Trustees, 1897 through 1955.

Sheridan Shore Yacht Club. sheridanshoreyachtclub.com.

U.S. Coast Guard. Letter dated April 21, 2005, to the Illinois Department of Natural Resources.

wikipedia.org.

Wilmette Harbor Association. Supplemental Submission to Metropolitan Water Reclamation District, 2010.

Wilmette Harbor Association. Letter dated February 11, 2013, to the Metropolitan Water Reclamation District of Greater Chicago.

Wilmette Historical Museum. wilmettehistory.org.

Wilmette Public Library. Wilmette History Collection. Index of historic street names.

Table 1: North Shore Channel Contract Sections

Section	Length in Feet	Upstream End Nearest (1)	Downstream End Nearest (1)	Constructed By
1	3,250	Lakefront	Jenks	District
2	4,080	Jenks	Green Bay (2)	District
3	1,480	Green Bay	Dodge	Schnable & Quinn
4	5,940	Dodge	Lake	Heyworth
5	2,630	Lake	Greenleaf	Heyworth
6	1,350	Greenleaf	Main	Fanning
7	5,330	Main	Howard	Healey (3)
8	4,730	Howard	Farwell	National Brick (4)
8A (5)	2,620	Howard	Touhy	Heyworth
8B (5)	2,110	Touhy	Farwell	National Contracting
9	5,100	Farwell	Glenlake	Heyworth
10 & 11	4,960	Glenlake	Summerdale	District (6)
12	4,010	Summerdale	Lawrence	Heyworth
Total	42,860			

Notes:

1. Street indicated may have to be extended to intersect with the North Shore Channel.
2. Green Bay Road formerly West Railroad Avenue.
3. Contract forfeited and district finished the work. The Healey contract excluded the short length enveloping the Oakton Street and C&NW railroad bridges and channel. This excluded work was performed partly under contract by Fanning.
4. National Brick was only responsible for excavation of a specific quantity. The contractors for sections 8A and 8B finished all remaining work.
5. Sections 8A and 8B subdivided section 8.
6. For unexplained reasons, sections 10 and 11 were combined as one section.

Table 2: North Shore Channel Road Bridge Construction

Bridge	Superstructure Fabrication and Delivery			Superstructure Erection and Completion			Open for Traffic
	Contractor	Start	End	Contractor	Start	End	
Argyle	W.A. Jackson	May 1909	March 1910	Heyworth	Dec. 1908	Aug. 1910	Aug. 1910
Brown	Toledo-Massillon	June 1908	June 1909	Heyworth	June 1908	May 1910	Sept. 1909
Bryn Mawr	Kenwood Bridge	July 1909	Oct. 1910	Roemheld	June 1910	Nov. 1910	Dec. 1910
Central	Indiana Bridge	June 1908	Oct. 1908	Kelly-Atkinson	Oct. 1908	May 1909	May 1909
Church	Toledo-Massillon	June 1908	June 1909	Heyworth	June 1908	May 1910	Feb. 1909
Dempster	Toledo-Massillon	June 1908	June 1909	Heyworth	Oct. 1908	April 1910	Feb. 1910
Devon	W.A. Jackson	May 1909	March 1910	Heyworth	March 1909	Aug. 1910	Aug. 1910
Emerson	Wisconsin Bridge	June 1908	July 1908	Heyworth	Sept. 1908	May 1910	Sept. 1909
Foster	W.A. Jackson	May 1909	March 1910	Heyworth	Dec. 1908	Aug. 1910	July 1910
Hill	Toledo-Massillon	June 1908	June 1909	Kelly-Atkinson	Oct. 1908	May 1909	Feb. 1909
Howard	W.A. Jackson	May 1909	March 1910	Roemheld	June 1910	Feb. 1911	Sept. 1910
Isabella	Toledo-Massillon	June 1908	June 1909	Volkmann	Sept. 1909	Jan. 1911	Jan. 1910
Lawrence	Used existing bridge	—	—	Heyworth	Raised existing bridge		Nov. 1910
Lincoln Ave.	Toledo-Massillon	June 1908	May 1910	Roemheld	June 1910	Feb. 1911	Nov. 1910
Lincoln St.	Toledo-Massillon	June 1908	June 1909	Volkmann	Sept. 1909	Jan. 1911	Nov. 1910
Linden	Toledo-Massillon	June 1908	June 1909	Kelly-Atkinson	Oct. 1908	May 1909	Feb. 1909
Main	Toledo-Massillon	June 1908	June 1909	Roemheld	June 1910	Feb. 1911	Nov. 1910
Oakton	Toledo-Massillon	June 1908	June 1909	Roemheld	June 1910	Feb. 1911	Sept. 1910

Table 2: North Shore Channel Road Bridge Construction (continued)

Bridge	Superstructure Fabrication and Delivery			Superstructure Erection and Completion			Open for Traffic
	Contractor	Start	End	Contractor	Start	End	
Peterson	W.A. Jackson	May 1909	March 1910	Roemheld	June 1910	Aug. 1910	Sept. 1910
Touhy	Toledo-Massillon	June 1908	June 1909	Kelly-Atkinson	April 1909	Aug. 1909	Sept. 1909
West Railroad	Fort Pitt Bridge	July 1909	May 1910	Schnable & Quinn	Mar. 1910	Nov. 1910	Nov. 1910

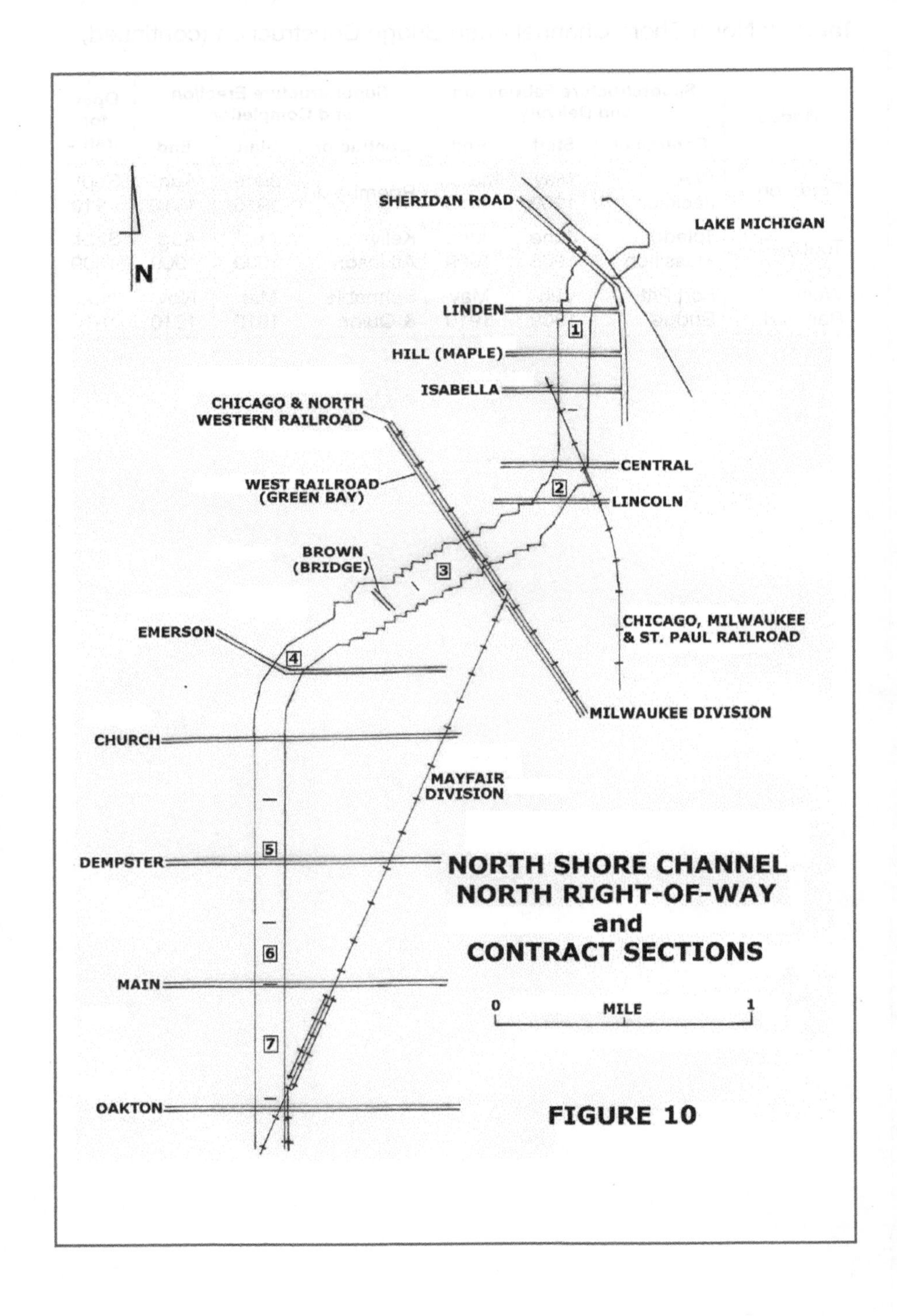

NORTH SHORE CHANNEL
NORTH RIGHT-OF-WAY
and
CONTRACT SECTIONS

FIGURE 10

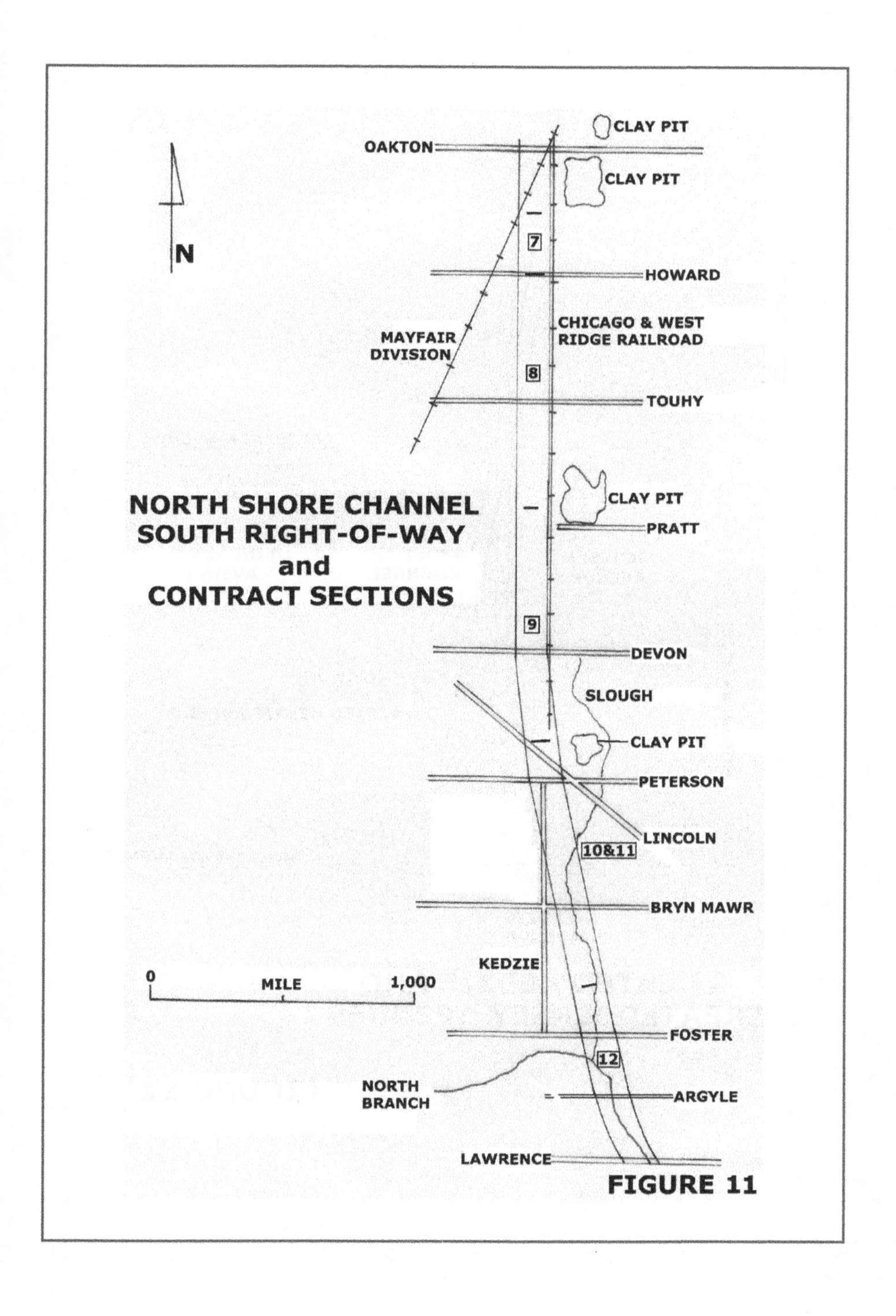
N
OAKTON
CLAY PIT
CLAY PIT
7
HOWARD
CHICAGO & WEST
RIDGE RAILROAD
MAYFAIR
DIVISION
8
TOUHY
CLAY PIT
PRATT
NORTH SHORE CHANNEL
SOUTH RIGHT-OF-WAY
and
CONTRACT SECTIONS
9
DEVON
SLOUGH
CLAY PIT
PETERSON
LINCOLN
10&11
BRYN MAWR
KEDZIE
0
MILE
1,000
FOSTER
12
NORTH
BRANCH
ARGYLE
LAWRENCE

FIGURE 11

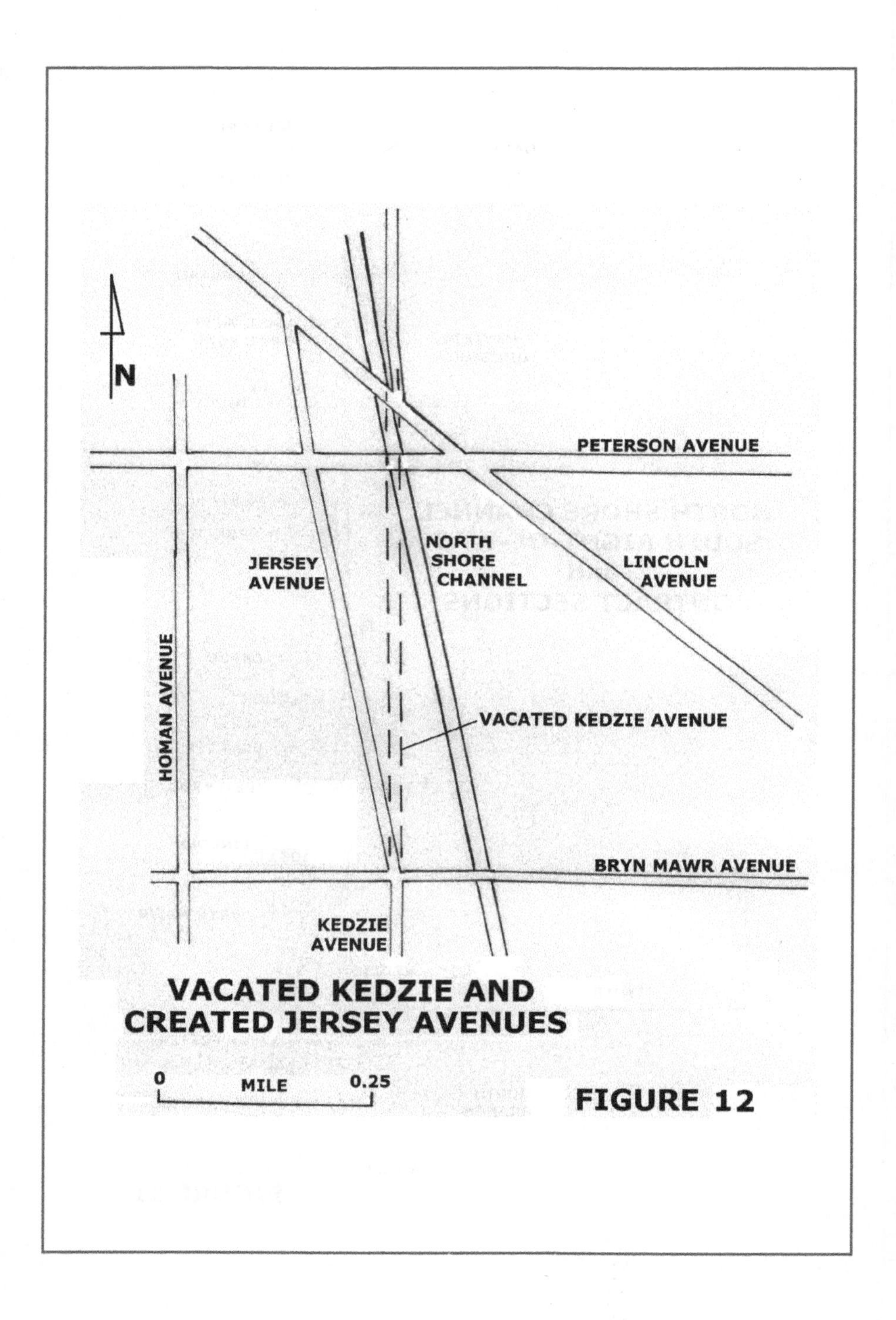
N
PETERSON AVENUE
JERSEY
AVENUE
NORTH
SHORE
CHANNEL
LINCOLN
AVENUE
HOMAN AVENUE
VACATED KEDZIE AVENUE
BRYN MAWR AVENUE
KEDZIE
AVENUE
VACATED KEDZIE AND
CREATED JERSEY AVENUES
0
MILE
0.25

FIGURE 12

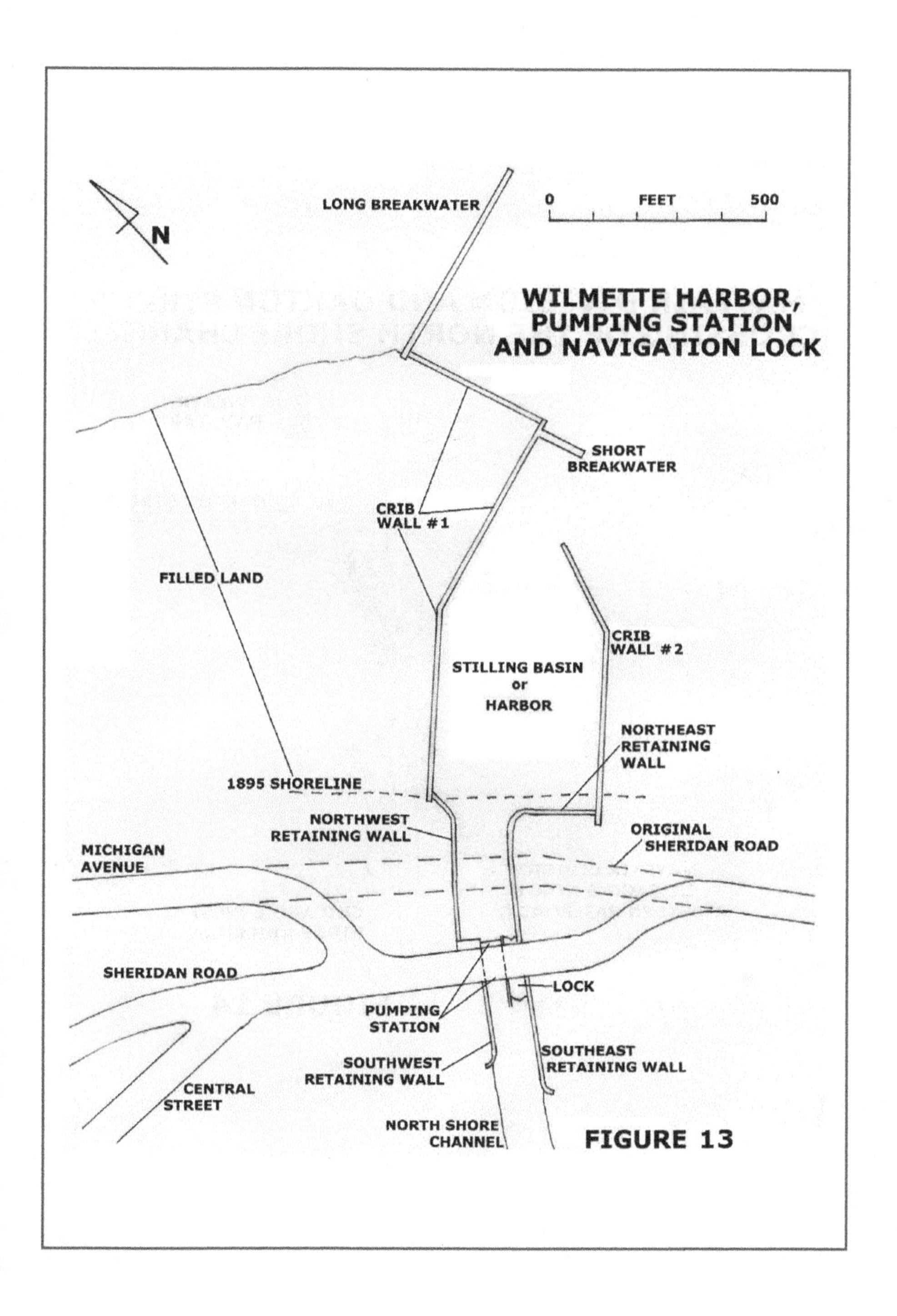
N
LONG BREAKWATER
0
FEET
500
WILMETTE HARBOR,
PUMPING STATION
AND NAVIGATION LOCK
SHORT
BREAKWATER
CRIB
WALL #1
FILLED LAND
CRIB
WALL #2
STILLING BASIN
or
HARBOR
NORTHEAST
RETAINING
WALL
1895 SHORELINE
NORTHWEST
RETAINING WALL
ORIGINAL
SHERIDAN ROAD
MICHIGAN
AVENUE
SHERIDAN ROAD
LOCK
PUMPING
STATION
SOUTHWEST
RETAINING WALL
SOUTHEAST
RETAINING WALL
CENTRAL
STREET
NORTH SHORE
CHANNEL

FIGURE 13

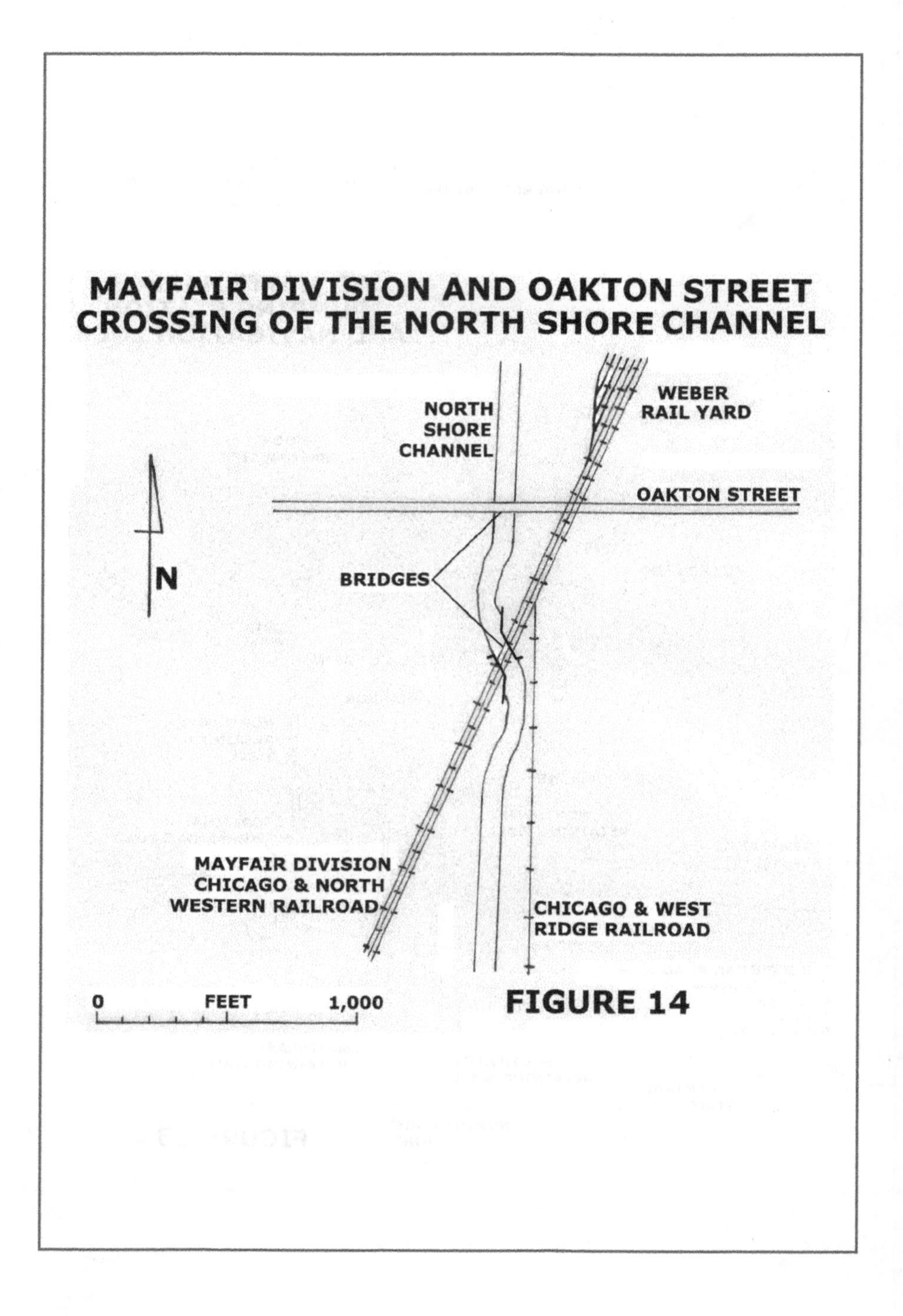
MAYFAIR DIVISION AND OAKTON STREET
CROSSING OF THE NORTH SHORE CHANNEL
WEBER
RAIL YARD
NORTH
SHORE
CHANNEL
OAKTON STREET
N
BRIDGES
MAYFAIR DIVISION
CHICAGO & NORTH
WESTERN RAILROAD
CHICAGO & WEST
RIDGE RAILROAD
0
FEET
1,000
FIGURE 14

Photograph 3.1: The National Brick Company buildings south of Howard Street are viewed from the west side of their clay pit looking southeast on July 6, 1907. The clay pit in the foreground excavated by National Brick will become part of the North Shore Channel. The railroad tracks and wooden incline are used to transport excavated clay into the building for making bricks. The long kiln shed in the left background continues to be used by multiple tenants to the present time. (MWRD photo 3762)

Photograph 3.2: After a wet, late winter, the flooded farm fields are viewed on March 6, 1908, between Pratt Boulevard and Touhy Avenues looking southwest from the North Shore Channel route. The property owners, sisters Mary E. and Julia P. Turner, lived in the house on the left. Wet farm fields were frequent throughout the area along the north-south route of the eventual channel. (MWRD photo 3861)

Photograph 3.3: Looking north from Oakton Street along the east line of recently acquired property for the channel shows flooded farm fields on April 19, 1909. The twin towers in the distance belong to the Fanning Company, a contractor excavating the channel near Main Street. The fence in the foreground marks the west right-of-way of the Chicago & North Western Railroad Mayfair Division Line, and the water tank and trestle on the right are part of the Weber rail yard north of Oakton Street. (MWRD photo 4079)

Photograph 3.4: Looking north from Foster Avenue in Chicago on April 19, 1909—is a flooded slough, the former and mostly filled in Big Ditch. Channel excavation had not yet begun in this area and the view shows the occasionally wooded areas along the channel route. The farm field to the right lies fallow. (MWRD photo 4075)

Photograph 3.5: Compared to Photograph 3.4, the slough north of Foster Avenue is not flooded on July 12, 1909. The trees are foliated and the farm field has been cultivated. The narrow water-filled channel is what is left of the Big Ditch. The farm field to the right may be part of the Budlong Pickle Company. (MWRD photo 4113)

Photograph 3.6: The sweeping shoreline of Lake Michigan is visible to the north from crib wall number one on May 9, 1908. The crib wall was built out into the lake adjacent to the cut in the bluff for the channel to contain spoil from channel excavation in Evanston and Wilmette. Few homes are located along what is presently Michigan Avenue at the top of the natural lakefront bluff. (MWRD photo 3863)

Photograph 3.7: The landfill is not only used for the disposal of excavated spoil, but also for the storage of materials, as seen in this view from the temporary Sheridan Road trestle on September 21, 1908. A dinky locomotive is pulling a string of dump cars around a loop of track after disposal of excavation spoil. Barges and work boats in the stilling basin are used for construction and maintenance of the crib walls. Crib wall number one extends into the lake in three segments, following a dog-leg pattern. (MWRD photo 3908)

Photograph 3.8: A dredge is at work in the stilling basin along with other work boats in this northeast view from the Sheridan Road trestle on September 8, 1909. Clay dredged from the bottom has been stockpiled pending loading onto scows for placement in the landfill or towed to a disposal area in the lake. Two sail boats are moored along crib wall number two, foretelling the future use of the stilling basin. (MWRD photo 4224)

Photograph 3.9: June 6, 1910. The lakefront landfill is still growing, but at a much reduced rate now that most channel excavation has been completed. Machines on the landfill are grading the surface. The temporary trestle for Sheridan Road, formerly at left, has been removed and traffic has been rerouted over the Linden Avenue Bridge. To the right of center, an earthen embankment keeps water in the stilling basin from the excavation for the pumping station, which is in progress north of the large tree, but out of sight below the higher ground. (MWRD photo 4389)

Photograph 3.10: A decade later, a young lad watches a hydraulic dredge suck up sediments adjacent to crib wall number one near the stilling basin inlet channel on September 12, 1920. Ten years after the landfill was completed, the stilling basin continues to need attention. To the southeast, the dredge pipeline stretches far beyond crib wall number two depositing spoil far from the inlet channel. (MWRD photo 7828)

Photograph 3.11: Laborers are constructing a concrete cap on top of the rock-filled timber crib wall number two on October 22, 1920. The contractor's floating plant consists of the materials and mixer barge at left and a work platform and derrick barge in the center. Wooden forms for the concrete cap have been fabricated and are laying on the work platform. Pilings have also been driven next to the crib wall to stabilize and protect it from boats. (MWRD photo 7891)

Photograph 3.12: A section of crib wall number one is being removed in this northeast view on October 22, 1920. Where the rock-filled timber crib wall was in disrepair and could not be stabilized to receive the new concrete cap, it was removed by the derrick on the barge using an orange-peel bucket and replaced with new pilings and concrete wall. (MWRD photo 7893)

Photograph 3.13: The timber crib between the two rows of piles is being filled with rock by the dipper dredge at the far end of the new long breakwater on October 22, 1920. A reinforced section of crib wall number one in the foreground connects to the land end of the long breakwater. The long breakwater was needed to blunt the force of storms from the north and northeast, and reduce the accumulation of sand in the stilling basin inlet. (MWRD photo 7894)

Photograph 3.14: December 5, 1920. The far end of the concrete cap in the foreground on crib wall number one has yet to be completed to connect with the short breakwater in the center. Out of view to the left is the crib wall number one connection to the new long breakwater observed in Photograph 3.13. (MWRD photo 7956)

Photograph 3.15: The stilling basin is looking more like a harbor on April 27, 1921, now that construction equipment has been removed. The northwest and northeast retaining walls appear in the left and right foregrounds, respectively. The new northwest harbor wall appears on the left at mid-depth and the concrete capped crib wall number two is across the harbor on the right. The inlet channel appears in the center background as it was built, and behind it the far end of the new 500-foot long breakwater appears in the background. The trees in the park on the left are ready to leaf out, and the lighting adds a touch of elegance. (MWRD photo 8125)

Photograph 3.16: Filling the new 500-foot breakwater with rocks has been completed on September 20, 1921. Later, this breakwater was reinforced with more piling on each side and surfaced with a concrete slab. It was popular for fishing and viewing of the shoreline. (MWRD photo 8422)

Photograph 3.17: This view from crib wall number one looking southwest shows the temporary trestle bridge for Sheridan Road spanning the cut in the bluff for the channel on May 9, 1909. A low embankment at the shoreline remains across the cut to separate the lake from excavation of the channel beyond the trestle. In the right foreground is rock-filled timber crib wall number one. Coming over the bluff at right is a string of dump cars pushed by a dinky locomotive transporting excavated spoil to be dumped in the landfill. (MWRD photo 3864)

Photograph 3.18: A dinky locomotive has brought four dump cars alongside the steam shovel to receive excavated spoil. When loaded, the locomotive will back down the slope and be switched to the track on the left, push the loaded dump cars up the track and onto the landfill across Sheridan Road. The early stages of construction of the channel attracted many observers on May 9, 1908. (MWRD photo 3870)

Photograph 3.19: The dragline excavator used near Emerson Street was owned by the James O. Heyworth Company, one of the more successful District contractors. The dragline was more efficient than a steam shovel in excavating the clay subsoil along the channel. After filling the bucket and pulling it to near the top of the cut at left, the operator lifts the bucket, swivels the machine, and tips the bucket to dump the load of clay onto the spoil pile, as shown on September 21, 1908. West of Railroad Avenue it was permissible to stockpile the spoil in piles along the channel route. (MWRD photo 3913)

Photograph 3.20: South of Main Street, another contractor, John T. Fanning Company, used this two-tower configuration for channel excavation. The operation is similar to a dragline, but employs two buckets operated separately from each tower, thus allowing excavation from both sides simultaneously. In this view on December 22, 1908, one bucket is being dumped behind the right tower and the other is being loaded in the cut. (MWRD photo 3977)

Photograph 3.21: James Heyworth brought in a larger dragline for excavation on Section 12, the reach from Lawrence Avenue to north of Foster Avenue. The bucket is shown on July 12, 1909, being pulled toward the machine against the slope, thus loading the bucket. Heyworth even gave his machines names, like *McCormick* and *Roosevelt*. (MWRD photo 4112)

Photograph 3.22: The District used steam shovels for channel excavation south of Peterson Avenue, shown here loading dump cars behind a dinky locomotive on December 23, 1908. The dump cars were towed to a nearby clay pit for spoil disposal. Besides being less efficient than draglines, the steam shovel worked in the cut rather than on the top of the channel bank. (MWRD photo 3989)

Photograph 3.23: The flood in April 1909 caused much havoc along the channel route. South of Bryn Mawr Avenue a steam shovel remained partially underwater on April 5, 1909, and the drainage ditch on the east side of the channel excavation was running bank full. It took several days of lost productivity to pump out the water and to repair the steam shovel before work could resume. (MWRD photo 4067)

Photograph 3.24: Looking southeast over construction of the North Branch Dam on January 20, 1910, concrete work has been completed and laborers are cleaning up construction debris. The dragline is preparing to remove the downstream berm isolating the work area of the dam from the channel. The Argyle Street Bridge east abutment and pier is in the left background. (MWRD photo 4281)

Photograph 3.25: Looking southwest at the downstream side of the dam on January 20, 1909. Backfill must be placed against the high side walls and downstream wing walls before the dam is complete. Five buttresses support the vertical overflow wall of the dam. The notch in the top of the far end of the vertical wall is for low flow. (MWRD photo 4282)

Photograph 3.26: The completed North Branch Dam on May 24, 1910, is passing a modest flood. The streamflow exceeds the capacity of the low flow spillway notch on the south end of the dam and is passing over the entire length of the overflow wall. Looking west from the east bank of the channel, the cupola atop Old Main is in the distance. Old Main was the first building on the new campus of North Park College and was the tallest building on the city's North Side at the time. (MWRD photo 4365)

Photograph 3.27: Near Dewey Avenue looking southwest at the transition from Section 3 to Section 4 on June 6, 1910. The Brown Street (later renamed Bridge Street) Bridge is in the distance and closer, the channel bottom is loaded with clay resulting from slope failures on both channel banks. Unstable slopes were prevalent from this point to the lakefront. (MWRD photo 4405)

Photograph 3.28: Looking north along Dewey Avenue and the timber flume for diversion of the Dewey Avenue sewer over the channel excavation on June 6, 1910. This sewer was a major outlet for the area north of the channel, and flowed south to Emerson Street and east to Lake Michigan. The size of the sewer precluded rerouting north of the excavation. (MWRD photo 4408)

Photograph 3.29: Slope failure and a landslide on August 27, 1910, wrecked the flume across the channel at Dewey Street resulting in sewage discharging into the excavation. This view looking north on August 29, 1910, shows total destruction of the flume. A Schnable & Quinn dragline at right is removing some of the fill in the channel. The unstable soil precluded rebuilding the flume, and sewage was allowed to flow into the unfinished channel since completion by the end of the year was expected. (MWRD photo 4498)

Photograph 3.30: After water was admitted to the channel late in November 1910 and pumping lake water began in April 1911, annual dredging was necessary to maintain the capacity of the channel until the slope stability problems were brought under control. A dipper dredge and scow are shown northeast of Emerson Street on December 12, 1913. Dredge spoil was often stockpiled in the stilling basin until it could be transported to Lake Michigan disposal areas. (MWRD photo 4893)

Photograph 3.31: A pile driving derrick and scow of timber sheeting works north of the Linden Avenue Bridge on the east bank of the channel on January 30, 1914, for a project to demonstrate a method of improving slope stability. Timber piles and sheeting are being driven vertically through the soft clay into deeper stiffer clay to provide lateral support for the soft clay. Later, the side slope will be cut back and armored with rock riprap to resist erosion. (MWRD photo 4905)

Photograph 3.32: A completed section of the slope stability demonstration project on the northwest bank of the channel west of West Railroad Avenue (later renamed Green Bay Road) on January 30, 1914. The demonstration proved effective and the method was used elsewhere. Notice the line of distinctive District transmission towers in the background, supplying electrical power to the Wilmette Pumping Station. (MWRD photo 4909)

Photograph 3.33: Hydraulic sluicing in action on the east bank north of Isabella Street on November 1, 1916, shows two laborers directing the nozzle stream at the base of the cut. A steam-driven pump on the barge supplies high pressure water. The loosened soil ends up in the channel, is removed by dredging, and towed in scows to a clay pit or the lake for disposal. Evanston Hospital is the large building right of center background. (MWRD photo 5651)

Photograph 3.34: Both channel side slopes in the reach between the Chicago, Milwaukee & St. Paul Railroad (now the Chicago Transit Authority Purple Line) and Isabella Street are being stabilized using hydraulic sluicing and the sluiced material in the channel is being removed by dredging on September 19, 1917. A golf course occupies some of the land on each side of the channel. (MWRD photo 6197)

Photograph 3.35: A damaged drainage ditch drop chute on the west side of the channel, north side of Dempster Street on April 11, 1919. Overflowing storm drainage has eroded the soils causing the concrete chute to collapse. The unstable clay side slopes and erosion was the cause of numerous similar failures along the channel route and became a continuing maintenance burden for the District until street drainage and sewer systems were installed. (MWRD photo 6972.5)

Photograph 3.36: The south drainage ditch drop chute on the west bank at Lincoln Avenue in Chicago has been rebuilt on April 11, 1919. The disturbed soil was removed; the void filled with broken rock, gravel, and broken concrete; a new concrete drop chute was cast; and the sides of the chute were paved with mortared riprap to resist erosion. (MWRD photo 6979)

Photograph 3.37: The drainage ditch drop chute on the west side of the channel and north side of Church Street was particularly bad on May 13, 1919. In addition to the drop chute, the bridge abutment was severely undermined endangering the safety of persons and vehicles crossing the bridge. (MWRD photo 7070)

Photograph 3.38: Erecting the distinctive District transmission tower near the southwest corner of the intersection of Colfax and Jackson Streets in Evanston for electrical service to the Wilmette Pumping Station on March 2, 1909. The transmission line paralleled the west side of the channel from Foster Street in Chicago to Sheridan Road in Wilmette. The electricity was generated at the District's Lockport hydroelectric powerhouse. (MWRD photo 4042)

Photograph 3.39: The transmission line began at the District's Leavitt Street Substation on Fullerton Avenue and ran underground beneath Elston and Kedzie Avenues to Montrose Avenue. It continued north as an overhead line in the alley west of Kedzie Avenue to Lawrence Avenue, east to Troy Avenue and north in Troy to the channel at Foster Avenue. On August 7, 1923, the line is shown through a congested shopping area along the north side of Lawrence Avenue looking east from the alley west of Kedzie Avenue. (MWRD photo 9932)

Photograph 3.40: Looking north on August 29, 1910, across the Wilmette Pumping Station site, pump tunnel formwork is to the right of center and farther right, a pile driver is driving foundation piles for the lock chamber out of view. The completed southwest retaining wall is to the left of center and the hole in the retaining wall is the outfall for the North Shore Intercepting Sewer, construction of which began in 1914. Above this wall is a temporary trestle for the delivery of construction materials and a concrete mixing plant. (MWRD photo 4485)

Photograph 3.41: Close up view of the downstream ends of the pump tunnel formwork on August 29, 1910. The small square holes are for the shafts connecting the screw pump propellers to the gear reduction units and electrical motor. The large holes in the foreground are the pump tunnel outlets, which will be in the west wall of the pumping station and fitted with flap gates. (MWRD photo 4490)

Photograph 3.42: Eleven days later, on September 9, 1910, the area lakeward of the pumping station is being excavated as close as possible to the embankment holding back the water in the stilling basin for construction of the northeast and northwest concrete retaining walls. The temporary trestle for Sheridan Road, which crossed this area where the steam shovel is working, has been removed. (MWRD photo 4515)

Photograph 3.43: Two months later, on November 2, 1910, the walls and windows of the pumping station are being installed and the flap gate frames on the tunnel discharge ports are in place. The small holes above the pump tunnel ports are for the flap gate counterweight chains. The concrete floor of the discharge channel remains to be finished, but the wall separating the pump discharge channel from the lock has been completed. The north span of the bridge must await removal of the concrete mixer. In the background, the earthen dike continues to hold back water in the stilling basin. (MWRD photo 4543)

Photograph 3.44: Three weeks later, on November 25, 1910, the east miter gates for the lock are in position. This view is from the stilling basin side of the lock. When swung open, each gate will be in a recess in the lock wall. Each gate is fitted with three small slide gates at the bottom to be used for emptying or filling the lock chamber to pass a boat through the lock. The miter gates point downstream since the higher water level will be in the channel when pumping occurs. (MWRD photo 4548)

Photograph 3.45: Looking from the west side of the channel southwest of the pumping station on November 25, 1910, laborers are seen completing the installation of the southwest miter gates for the lock. The pump tunnel flap gates and chains are already in place. In four days, water will be admitted to the channel in advance of the coming winter. The masonry walls of the pumping station are in place, but the windows have yet to be glazed. The erection of the Sheridan Road Bridge structural steel awaits completion. (MWRD photo 4552)

Photograph 3.46: The Wilmette Pumping Station and Sheridan Road Bridge viewed from the northwest bank of the channel southwest of the station in May 1916. Stairways leading down from the roadway and the public comfort station under the northwest bridge span have yet to be installed, but the pumps are in operation delivering the required dilution water to the channel. (MWRD photo 5472)

Photograph 3.47: The Chicago, Milwaukee & St. Paul Railroad (presently known as the Chicago Transit Authority Purple Line) tracks on the left in this north-facing view are being relocated temporarily on the diversion tracks on the right so a bridge can be constructed where the tracks pass over the channel. A temporary timber trestle bridge, center background, has been built over the channel for the diversion. In this September 1, 1908, view, the steam shovel is clearing away soil in the bottom of the excavation resulting from the slide of unstable soil to the right of the steam shovel. (MWRD photo 3892)

Photograph 3.48: The Milwaukee Electric Railroad used this bridge on the right-of-way owned by the Chicago, Milwaukee & St. Paul Railroad north of Central Street in Evanston. The through-truss and plate girder double-track bridge was completed and in service one year and seven months before this north view on April 24, 1911, but water has only been in the channel for five months. (MWRD photo 4616)

Photograph 3.49: Work is just getting underway on the Chicago & North Western Railroad Milwaukee Division (presently known as the Metra Union Pacific / North Line) crossing of the channel on April 26, 1909, with the completion of the track diversion to the right mainline tracks. The buildings on the right face West Railroad Avenue (later renamed Green Bay Road) near Lincoln Street in Evanston. The houses have been replaced by Haven Middle School and Kingsley Elementary School. (MWRD photo 4092)

Photograph 3.50: By August 15, 1910, the new steel plate girder spans for the Milwaukee Division and the West Railroad Avenue crossings have been completed. Channel excavation has also been completed and laborers are constructing the concrete struts between the bridge abutments. To allow the West Railroad Avenue Bridge to be built, train travel on the new railroad bridge had to be restored and the tracks for both approaches have been placed on timber trestles until the embankment could be built. (MWRD photo 4441)

Photograph 3.51: Looking southwest at the Milwaukee Division steel plate girder railroad bridge on April 24, 1911. The plate girder span has been completed as well as the approach embankment and tracks. The West Railroad Avenue Bridge is viewed in the opening under the railroad bridge. Water has been in the channel for five months and in the left foreground is evidence of side slope failure. (MWRD photo 4621)

Photograph 3.52: The east reinforced concrete abutment for the Chicago & North Western Railroad Mayfair Division Bridge is under construction in this north view on July 12, 1909. The tops of circular concrete piers reaching to bedrock are in the foreground and will be covered by an extension of the abutment. A concrete mixer is on top of the bank at right. The abutments are long in the direction of the channel because of the skewed crossing, about 20 degrees. Presently, this bridge remains, but the rail line no longer exists. (MWRD photo 4119)

Photograph 3.53: Both abutments are complete and structural steel erection has begun for the Mayfair Division Bridge. On September 30, 1909, a construction mishap occurred when the timber boom of the derrick lifting a girder failed, dropping the broken boom and girder into the channel between the two abutments. The remaining mast and one strut of the derrick are viewed at right on top of the east abutment. (MWRD photo 4240)

Photograph 3.54: Looking south from the Mayfair Division Bridge on November 25, 1910, the North Shore Channel will be filled with water in four days. The railroad bridge has been carrying trains for a year and the temporary trestle for the track diversion has been removed. In the left foreground part of the bridge framing appears spanning between the two abutments and supporting the track bed. The Howard Street Bridge appears in the distance. (MWRD photo 4566)

Photograph 3.55: Looking east along a strip of land on September 21, 1908, south of Central Street in Evanston that has had the topsoil removed prior to construction of a diversion for Central Street, which is barely visible at left along the telephone poles. Trolley tracks and pavement will be laid down for the diversion on land that was part of a golf course. In the center background, also barely visible in the haze is the Central Street railroad depot on the Milwaukee Electric Line. The tracks run on grade, but were elevated south of the railroad bridge prior to 1920. (MWRD photo 3903)

Photograph 3.56: A trolley heads west on the completed Central Street Bridge on April 26, 1909. This bridge, originally a three-span steel plate girder structure with steel cross beams, oak sub-floor, and creosote paving blocks, has been replaced with a modern concrete and steel bridge. Trolleys have been replaced with buses. The plume of steam at left is coming from a steam shovel excavating the channel underneath the south side of the bridge. (MWRD photo 4090)

Photograph 3.57: Oakton Street is not much more than a rutted wagon road in this west view on July 12, 1909. Laborers are excavating shafts for three of the four piers for the road bridge over the channel. The clay is dug by hand and hoisted by hand winch to the surface. The earthen sides of the shaft are supported by wood planks held by circular rings inside the shaft. The pier is formed by filling the shaft with reinforced concrete. This construction was typical for all bridges using piers to bedrock or hardpan clay for foundations. (MWRD photo 4121)

Photograph 3.58: District laborers are constructing the concrete strut between the two opposing north piers on either side of the channel under the Linden Avenue Bridge on August 15, 1910. The strut between the two south piers has been completed. The strut resists the lateral pressure of the channel side slope on the vertical pier and without them the piers might move toward the center of the channel. This bridge is another example of a steel plate girder road bridge. Note the deep cut of the channel and the higher surrounding land, allowing the two abutments to be spread footings founded on undisturbed soil. (MWRD photo 4487)

Photograph 3.59: A stiffleg derrick is used to lift and place structural steel members in erecting the Emerson Street Bridge on April 26, 1909. Both structural steel bents have been placed on the tops of the piers and the west approach span is being assembled. The next girder to be lifted into place lies on the channel bank under the west approach. At left is the temporary trestle bridge for the Emerson Street diversion. (MWRD photo 4093)

Photograph 3.60: Erecting the superstructure steel for the truss-type Argyle Street Bridge on May 24, 1910, using a stiffleg derrick. This north view shows the bents on top of the substructure piers are reinforced concrete construction. The surrounding land is floodplain; hence the high abutments also founded on deep piers to hardpan clay. In the background behind the east bent of the bridge is the North Branch Dam. (MWRD photo 4364)

Photograph 3.61: Laborers are tearing up the wooden deck of the Oakton Street Bridge on October 14, 1921. Before the use of concrete for bridge decks, wood was the material of choice, but it only had a five- to ten-year life, requiring more frequent replacement. In the left background to the west is a building, elevated water tank, and gas holding tank, part of the Public Service Company of Northern Illinois manufactured gas plant on the south side of Oakton Street built in 1910. (MWRD photo 8509.5)

Photograph 3.62: View of the channel looking northeast from near Emerson Street on May 3, 1911, with the Brown Street Bridge in the background. Water has been flowing in the channel for less than a month. Spoil piles were prevalent along the channel west and south of West Railroad Avenue. Most of this spoil was eventually removed by the National Brick Company and used in the manufacture of bricks. (MWRD photo 4645)

Photograph 3.63: View looking north from the east side of the Argyle Street Bridge showing the District's work boat in the foreground, and the North Branch Dam and Foster Avenue Bridge in the background on May 9, 1911. A person is walking along the path west of the channel and water is flowing under the bridge from the dam and the Wilmette Pumping Station. Only one sewer was known to have been discharging to the channel as of this time. (MWRD photo 4651)

Photograph 3.64: Shuttered dinky locomotives on the rail siding south of Hill Street (later renamed Maple Avenue) in Wilmette on May 8, 1913. Storage of District equipment and unused construction material along the channel right-of-way several years following completion of channel construction drew complaints from neighbors and the village. Some of the equipment was transferred to and used on the Calumet-Sag Channel and the area wasn't cleaned up until near the end of the decade. (MWRD photo 4816)

Photograph 3.65: Wilmette Harbor in a 1938 aerial photograph appears the same as constructed by the District. Ownership of most of the landfill, Gilson Park, was transferred to the Wilmette Park District by the state in 1911. The District retained ownership of a small area on each side of the harbor and has been leased. Many boats are using the harbor. (MWRD aerial photo archives)

Photograph 3.66: Wilmette Harbor in a 2009 aerial photograph appears much different with the extension of the inlet channel into the lake, significant accretion of sand both north and south of the inlet channel, and many more boats. (Wilmette Harbor Association photo)

Photograph 3.67: The Wilmette Pumping Station was rehabilitated in 2012 and 2013 by the District, and this November 5, 2014, view of the harbor side shows the rebuilt intake structure with four sluice gate motor operators above the four water tunnels. At left is the former lock chamber, now used as another water passage. (Photo by the author)

Photograph 3.68: The channel side of the Wilmette Pumping Station on July 4, 2014, shows four sluice gate motor operators above the four water tunnels from left to center. Sluice gates at both ends of the water tunnels allow dewatering of the tunnels for maintenance. The two tunnels at left contain pumps; the pumps have been removed in the two tunnels to the right. Farther right and higher at street level are three sluice gate motor operators above the three new sluice gates for controlling water flow through the old lock chamber. (Photo by the author)

Chapter 4

North Area Sewersheds and Watersheds

Introduction

The *dilution system* refers to the District's canals constructed to use water from Lake Michigan to dilute and flush sewage to the Des Plaines River at Lockport; it also refers to intercepting sewers, outlet sewers, and pumping stations to convey sewage from communities to the canals. The term was in use from the time the Sanitary & Ship Canal was opened early in 1900. Its use ended in August 1919 when the trustees adopted a program for the construction and operation, over a 25-year period, of sewage treatment plants with the goal of reducing by 50% the waste load in the Sanitary & Ship Canal discharged to the Des Plaines River. Subsequent to 1919, the *sewage treatment system* referred to the work of the District and included the construction of intercepting sewers to collect and convey sewage as well as treatment plants to reduce the load of solids discharged to the canals. What had been built for the dilution system was gradually incorporated into the sewage treatment system.

In 1900, proven sewage treatment technology for large urban populations was nonexistent. With dilution, sewage would be less offensive and its treatment by the natural process of a flowing channel or river would be enhanced and hastened. The dilution system was a success. Inherent in the dilution system was the elimination of sewage discharged to the lake, so the system protected the quality of Lake Michigan for water supply for the growing metropolitan area. However, due to population exceeding expectations and to the large amount of industrial waste discharged to the canals, the dilution system was less than successful for the canal system and downstream rivers. More dilution water may have compensated for the greater population, but the District was under pressure from the federal government to not divert as much water from Lake Michigan as allowed under state law. In addition, the Canadian government and other states around the Great Lakes were objecting to the quantity of diversion.

The waste load from industrial development was not considered or included in the dilution ratio required in the Act of 1889. Hence, this waste load could not legally be compensated for by more dilution. The District, through persuasion and even threats of litigation with the industries, attempted to deal with the waste loads but discharging industrial waste to rivers was not unlawful at the time so litigation would not be successful. As a result of this situation and similar experience in other cities in the U.S. and Europe, technologies for treatment of sewage were being studied widely. The District began its own experiments in 1909; ten years later, with the results of its own experiments and the work of others, the *sewage treatment system* was established. District experiments included aerobic and anaerobic biological treatment on human sewage and trade wastes. The District also built one small treatment plant for human sewage before 1919. The 1919 decision set the course of the District in a new direction.

Dilution System Outlet Sewers

In 1911, with the improved North Branch and the new North Shore Channel completed and partially in operation, a proper outlet for sewage and stormwater for the city's North Side and northern suburbs became available. As discussed earlier, the City and District cooperated on the North Side of the city in eliminating the discharge of sewers to the lake. In anticipation of the construction of an intercepting sewer for North Shore suburbs, the southwest retaining wall at the Wilmette Pumping Station was built with a sewer outfall.

After 1910, Evanston and other North Shore suburbs requested the District to construct intercepting sewers discharging to the canal and eliminate sewers discharging to the lake. However, even though the attorney had issued an opinion in 1909 that sewers are clearly included under the District's authority for drainage, some trustees believed the District lacked the authority to build the sewers. The relationship between the District and North Shore suburbs became hostile. In addition, construction of Morton Grove village sewers, which discharged to the North Branch far upstream of the canal ignited citizen protest in that municipality and in others along the North Branch and West Fork. Litigation against the District was being threatened.

Tensions were eased and the course was set in February 1912 with the trustees agreeing to have plans prepared for an intercepting sewer to serve the North Shore suburbs. Starting in 1913, the North Shore Intercepting Sewer was built from just west of the Wilmette Pumping Station; it followed Sheridan Road north and west to Cherry Street in Winnetka. Under other streets and rights-of-ways, the intercepting sewer was extended to include Glencoe, and by 1916, most sewers discharging to the lake from the four lakefront suburbs had been intercepted, with all sewage discharged into the new canal. Glencoe's municipal septic tank was also connected. At the time it was too far to build a sewer to serve Morton Grove; instead a treatment plant was built as explained later.

To drain neighborhoods along the canal, the City and Evanston proceeded with more sewer construction and sewer improvements, and the District cooperated in issuing permits to cross the channel right-of-way for the construction of sewers and outfalls. In May 1913, Evanston annexed areas south of Crain Street, and development required sewers. In April 1916, Evanston was granted permits to build three five-foot diameter sewers and outfalls at Cleveland, Greenleaf, and Mulford Streets which discharged to the canal. A year later another permit allowed Evanston to construct a six-foot diameter sewer discharging at Emerson Street.

East of Chicago Avenue, Evanston presented a more difficult problem with several sewers discharging to the lake. Similar to what was done in the city and with the cooperation of Evanston, a plan was developed for two intercepting sewers along the lakefront, one from the south city border draining north and the other from near Emerson Street draining south. Together these two sewers intercepted the sewage at five locations from sewers discharging to the lake. At Lake Street, the two lakefront intercepting sewers converged and the sewage was conveyed west, to the Evanston Pumping Station located on the northeast corner at Elmwood Street. Using a ten-foot diameter intercepting sewer along Lake Street to the canal, the sewage and wet weather flow was lifted by pumps to allow flow westward by gravity. The pumping station had three dry weather centrifugal pumps rated at 4,300 gpm each and three wet weather centrifugal pumps rated at 15,000 gpm each. Originally the station was manually operated; presently it is automated and controlled from the O'Brien plant.

Between Chicago and Ridge Avenues in south Evanston, another low area, sewage from the intervening wide and subtle swale was collected in an intercepting sewer along Custer and Elmwood Avenues; it flowed from Howard Street north to Lake Street where an intercepting sewer connected to the ten-foot diameter sewer under Lake Street. A third intercepting sewer in Sherman Avenue drained the downtown area, flowing south and connecting to the Lake Street Sewer. The entire system of intercepting sewers and the pumping station was completed by 1920. In 1928 the Lake Street intercepting sewer was connected to the North Side intercepting sewer through an inverted siphon under the canal, and the sewage was conveyed to the North Side Sewage Treatment Works at Howard Street, recently renamed the O'Brien Water Reclamation Plant. An outfall at Lake Street on the east side of the canal allowed excess wet weather sewage and stormwater to be discharged directly to the canal.

Niles Center, later renamed Skokie, was sited upon a subtle watershed divide between the vast marshy expanse which approached Evanston to the east and the North Branch to the west. With poor drainage in the flat landscape, and being nearly equidistant from the North Branch and the new canal, in 1917 a two-mile-long outlet sewer running east along Oakton Street was built enabling elimination of sewage discharge to the North Branch. It was preferable to discharge to the canal because the channel was lower in elevation and designed to accept sewage. Discharge to the west, with its higher elevation and potential of fouling the stream, would be more problematic. In 1928 with the opening of the plant on Howard Street, the Niles Center Outlet Sewer was eventually connected to the North Side intercepting sewer.

Farther south the city's North Side, already extensively sewered by the City, needed no outlet sewers built by the District. Earlier sewers constructed by the City department of public works drained from the east and west to the canal. As neighborhoods developed north of Lawrence Avenue, the canal was available to receive sewers. Farther northwest, developing neighborhoods were served by City sewers along Bryn Mawr, Devon, Foster, Peterson and Lawrence Avenues, with diagonal streets, conveying sewage and stormwater to the canal.

North Side Intercepting Sewer

Following the 1919 sewage treatment decision, the plan for the north area included a treatment plant at Howard Street and a network of intercepting sewers to collect sewage from Fullerton Avenue on the south, the lakefront on the east and the county border on the north. The North Side intercepting sewer serves the north area with two branches, one north of Howard Street and the other south. Construction of the north branch from near Sheridan Road next to the Wilmette Pumping Station started in 1921. This branch follows the canal west and south along the west side of the canal to near Oakton Street. Between Green Bay Road and Howard Street it follows McCormick Boulevard.

Construction of the south branch began in 1925 at Howard Street. From Fullerton Avenue east of Elston Avenue on the south end, the intercepting sewer follows Fullerton, Elston, California, and Manor Avenues northward. To avoid the complex intersection of Belmont/California/Elston, the intercepting sewer follows Washtenaw Avenue and Melrose Street. At Giddings Street, the intercepting sewer turns north, crossing Lawrence Avenue to the west side of the canal, then follows the west side of the canal to Devon Avenue. North of Devon it follows McCormick Boulevard. Several different contractors built the intercepting sewer, which was divided into nine segments, as well as the local sewer connections.

The alignment of both north and south branches on the west side of the canal facilitated the connection of numerous municipal sewers draining from west to east. Numerous City sewers, already in place, were connected as part of the construction. Generally, a local sewer is connected to an intercepting sewer at a control chamber where a low dam in the local sewer deflects dry weather sewage to the intercepting sewer below. By diverting excess flow to the canal through an outfall, the control also protects the intercepting sewer from being overloaded during wet weather. Since the invert of most local sewers are below wet weather water levels in the canal, a tide gate in the local sewer prevents canal water from backing into the sewer and intercepting sewer. (See Figure 15.)

The north and south branches of the North Side Intercepting Sewer converge at Howard Street and enter the O'Brien plant northwest of the intersection with McCormick Boulevard. The sewage in the approximately 40-foot-deep sewer is lifted by huge pumps in the adjacent building to ground level as it starts through the treatment or reclamation process. That is discussed in more detail in Appendix A-9.

Sewage flowing in local sewers from east to west on the east side of the canal is conveyed to the intercepting sewer through inverted siphons under the canal. These siphons are usually built with two low flow pipes and one high flow pipe, the former for dry weather and the latter for wet weather. Separate shafts on both the upstream and downstream sides provide for access to clean-outs for maintenance. (See Figure 16.)

In four locations east of the canal, the flow of sewage under the canal is aided by pumping stations. The Evanston Pumping Station was discussed earlier as part of the Lake Street intercepting sewer. The North Branch Pumping Station was discussed in Chapter 3. Farther south, the Wellington Avenue Pumping Station on Clybourn Avenue, built in 1960, discharges sewage to the intercepting sewer through an inverted siphon under the canal at Barry Avenue. This station went into service in 1960 and drains an area of about 480 acres bounded by Roscoe Street and Ashland, Diversey, and Western Avenues. The station has three centrifugal pumps, for both sewage and stormwater, each rated at 7,500 gpm. A wet weather outfall is located near the siphon.

The Wilmette Pumping Station does not pump sewage, only lake water, but there is a small sewage pumping station located adjacent to the Wilmette Pumping Station. Named the Wilmette Lift Station to distinguish it from its larger namesake, it was placed in service in 1938. Its purpose is to pump sewage to the North Side intercepting sewer on the opposite side of the canal from a small area of about 100 acres of combined and separate sewers bounded by the canal, lakefront, and Isabella Street. For dry weather flow, the lift station has one pump rated at 1,200 gpm; for stormwater it has an overflow drop shaft to Deep Tunnel.

In addition, numerous short sewers and control chambers connect to trunk sewers located east of the canal, lead to inverted siphons to convey dry weather flow to the North Side intercepting sewer. For stormwater, at least one drop shaft to Deep Tunnel is associated with each inverted siphon.

Another important detail pertains to the *residual area*, the space from the intercepting sewer and control chamber to the canal. Since it was not practical to construct the North Side intercepting sewer on private property immediately adjacent to the canal, all drainage from the residual area is collected and conveyed to the intercepting sewer. Thus there are numerous short connecting intercepting sewers serving this purpose in thoroughfares and side streets.

O'Brien Water Reclamation Plant

Construction of the North Side Sewage Treatment Works began in 1923 at the Howard Street site. That site was within the municipality of Niles Center, presently named Skokie, but distant from developed areas. Two important factors were rail and water access. Rail was needed for delivery of construction materials and equipment and for delivery of chemicals and fuel for plant operation. Water access was needed for discharge of the treated sewage effluent. In addition, the site at Howard Street was partially owned by Clara F. Bass, a nonresident from whom the District had successfully purchased channel land near Bryn Mawr Avenue and Oakton Street. The North Side Sewage Treatment Works, completed and placed in service in October 1928, was renamed the O'Brien Water Reclamation Plant in 2012. Additional details are found in Appendix A-9.

Plant construction was accomplished through several different, competitively-bid contracts. Typical for the District at the time, pieces of electrical and mechanical equipment were purchased separately through furnish, deliver, and install contracts. The John Griffith & Son Company was the successful low bidder and principal construction contractor under five contracts for the: pump and blower building;

grit chambers and primary settling tanks; aeration tanks, operating galleries, and final settling tanks; main and grit buildings; and service building and fuel tanks.

It is helpful to use modern terminology for these plants, because what was termed a *sewage treatment plant* is presently referred to as a *water reclamation plant*. The modern terminology is more precise: The convenience and availability of water is being used as a medium to convey excrement, urine, and other waste—in aggregate called sewage—to a place where the water can be reclaimed or recovered from the waste and safely returned to the environment for reuse. *Water resource recovery facility* is another term gaining popularity. Sewage treatment may have been appropriate for its day, but today this public works service industry is not just recovering water but is also generating two principal byproducts—energy and fertilizer nutrients.

Following the plant opening in October 1928, bulkheads in the control chambers all along the North Side intercepting sewer from Sheridan Road in Wilmette to Fullerton Avenue in Chicago were removed, allowing sewage to flow to the treatment plant. Sewage was no longer discharged directly to the canal during dry weather.

More Intercepting Sewers

Within a decade following the opening of the plant, the intercepting sewer network was being expanded to provide additional capacity: for lakefront suburbs in 1935, to serve Northfield in 1939, and for Glenview and Golf in 1940. The southern part of Niles was served by the Niles Pumping Station, which was placed in service in July 1938. The station was located on the east side of Milwaukee Avenue a few blocks north of Devon Avenue. The next expansions had to wait until after World War II.

New residential areas in the western parts of the four lakefront suburbs were served by a new leg of the North Shore intercepting sewer built in 1947 along Green Bay Road from the North Side intercepting sewer to Glencoe; it relieved local sewers as well as the original intercepting sewer built in 1916. Lincolnwood was served by connections in 1947 and by extensions in 1956. Growing north suburbs away from the lakefront were served by the Howard Street intercepting sewer, with construction beginning in 1953 at the treatment plant and extending west along Howard Street. Branches of the intercepting sewer followed Austin Avenue, Ballard Road, Golf Road, Harms Road, Milwaukee Avenue, Sunset Ridge Road, Wagner Road, and many others serving Glenview, Golf, Morton Grove, Niles, Northbrook, and unincorporated areas. This expansion lasted into the 1970s.

The extension of the Howard Street intercepting sewer allowed the Niles Pumping Station to be removed from service in 1957; the station was demolished in 1985. The Walters Road Pumping Station on Walters Avenue west of Saunders Road in Northbrook was placed in service in July 1963 with three centrifugal pumps, each rated at 3,200 gpm. The station, which served an area east of the Des Plaines River that is approximately seven square miles, mostly residential and low, lifts the sewage into the gravity intercepting sewer about three-fourths of a mile to the east of the pumping station.

As areas developed and as inadequately sized sewers needed to be replaced in the early part of the twentieth century, the City continued building sewers. However, the Depression in the 1930s and World War II in the early 1940s slowed the program. Following the war, the City began a robust program of building relief sewers—sewers to relieve overloaded and under sized combined sewers—providing additional capacity for stormwater. Generally, there were older sewer outfalls under the bridges of thoroughfares on both banks of the canal. Relief sewers were constructed between the major thoroughfares spaced at generally half-mile increments, with large outfalls along the west side of the canal at Lawrence and Berteau Avenues, Grace and Roscoe Streets, and Logan Boulevard. These relief sewers drained large developed areas of the city to the west. From the east, major relief sewers were found at Addison, Devon, and Pratt.

The West Lawrence Avenue relief sewer was built in the early 1970s as a rock tunnel with a pumping station. The outfall was on the west bank immediately north of the Lawrence Avenue Bridge. The outfall, built prior to the District's Deep Tunnel, was eventually connected to the Deep Tunnel, thus eliminating the need for the separate pumping station. Many of the relief sewers are interconnected, providing the opportunity for stormwater to seek the path of least resistance to an outlet. Interconnection provides efficient drainage since storm rainfall is rarely distributed uniformly.

Today, in addition to the above, each municipality maintains its own sewer collection system. Since there is a sewer in virtually every street, the aggregate total length of sewers approaches 1,500 miles for the area tributary to the O'Brien plant. This vast underground infrastructure is not visible, but it works at all times to effectively drain the north urban area. And if something malfunctions, there are people who actually know all the sewers and how they work, and who can wake from a deep sleep and know where to go and what to do to make corrections. These people are the professional and technical staffs of the District and municipal public works or utilities departments.

Watersheds and Sewersheds

Flying into O'Hare or Midway Airports during night, coming in over Lake Michigan, all is darkness below. Then the northeast Illinois lakefront appears, a sharp edge of light. Soon, over the land below, one sees a massive rectangular grid defined by street lights and interrupted with diagonal expressways, railroads, and thoroughfares. The grid rolls out over the vast urban and suburban area, interrupted by blocks of darkness marking a cemetery, a forest preserve, a park, or a school campus. A few jagged or sinuous dark strips define canals or rivers. The grid is the structure of urban living, forcing otherwise free-roaming humans into order and place. You live over here, work over there, move along these routes, etc.

The grid makes urban existence livable since humans feed on communications, energy, and water, typical utility services that flow within the grid. And for what is cast aside, refuse trucks glide through alleys and streets, sewage and stormwater, out-of-sight, flows through drains and sewers beneath the streets. In flat topography like the Chicago area, the grid is usually rectangular; in hilly or rolling topography, the grid can have interconnected, polygon-like shapes.

In a natural landscape, the drainage network of swales, creeks, and streams are contained within a watershed, which is a region of land that drains to a common low point in a river or stream. The watershed divide is the highest land elevation surrounding the watershed, separating it from neighboring watersheds. All is above ground and visible. In wet weather, rainfall clings to trees and ground vegetation and soaks the land surface. A good amount will infiltrate the soil and percolate downward, nourishing vegetation and replenishing groundwater. Excess rainfall fills depressions and flows over the land surface, gathering in swales. Perhaps dry most of the time, in wet weather, swales become ephemeral streams, joining together and eventually becoming an overflowing perennial stream. Floodplains border the perennial stream floodway to provide temporary storage for floodwater. (See Figure 17.)

In the urban landscape, drainage is defined by its sewer system: it follows the grid, it is contained within sewersheds, it is below ground and not visible. Unlike a watershed, a sewershed is human-made, designed by hydrologists and engineers. The sewershed divide is a high elevation in the grid, but can also be arbitrarily set at other borders such as a municipal boundary, major thoroughfare, or subdivision boundary. Because of impervious surfaces such as roofs and pavements, during wet weather much less of the rainfall soaks into the ground or fills surface depressions in a sewershed. Perhaps only 0.1 inch is detained. Turf in lawns, parks, school yards, etc. may appear permeable, but the soil is packed tight by feet and turf maintenance equipment. Turf roots don't penetrate deep into the soil as native vegetation does.

Excess rain flows over the impervious and semi-pervious constructed surfaces into a street drainage inlet, then on to a combined or storm sewer. Hence the volume of rainfall runoff is greater and it reaches the sewer and sewershed outlet faster than in a natural watershed of comparable size. Urban sewers also contain sewage from homes and businesses, conveyed by water whose source is usually a public water supply that comes through a tap. In dry weather, used water disappears down the drain and flows through the sewer network to a treatment plant, but can overflow to a river when mixed with larger quantities of stormwater.

Sewers are either privately or publicly owned. A *private* sewer in a network is referred to as *lateral*; it conveys drainage from a building or property to a public sewer. Lateral sewers are usually six inches in diameter, but can be larger if necessary, and are the responsibility of the property owner. Generally, this responsibility extends to or beyond the property line to the connection to the public sewer under the street or in an easement; the practice varies by municipality. *Public* sewers are usually owned by municipalities, but can also be owned by a publicly regulated private utility.

Combined or separate are the two types of sewers, and a sewershed may contain either one or both. Public sewers are used in one of three ways: those that convey only sewage are identified as *sanitary* sewers, only stormwater are *storm* sewers, and both sewage and stormwater are *combined* sewers. Combined sewers are outdated and are no longer built, although many old combined sewer systems are still in use. In separate sewer systems, common in cities and towns developed since the 1950s, sewage and stormwater are collected and conveyed in separate pipe networks. The storm sewers will discharge to a waterway and the sanitary sewers will discharge to a sewage treatment or water reclamation plant. (See Figures 18 and 19.)

Sanitary sewers are not designed to overflow where sewage would reach a storm sewer or stream. These overflows are prohibited by the Clean Water Act. However, overflows do occur through accident or malfunctioning infrastructure and responsible utilities must have an emergency response plan to mitigate such occurrences and must take corrective action to prevent a reoccurrence.

The public sewer may be a branch sewer if it originates in the block(s) served by the branch sewer or it may be a trunk sewer if branch sewers feed into it at street intersections or easements. Whichever it is, within a city block there are numerous private laterals and several street stormwater inlets contributing flow to the public sewer, whether the sewer be combined, separate, or storm. Branch sewers range in size, but are usually 12 inches in diameter. As trunk sewers gather laterals and other trunk sewers, their size increases, diameters of 24, 36, or 48 inches are common. The network of sewers eventually reaches an outlet.

In combined sewer systems, typical of older cities where sewage and stormwater flow in the same pipe, the outlet will be an intercepting sewer. The intercepting sewer collects and conveys dry weather sewage and wet weather flows to a sewage treatment or water reclamation plant. Excess combined sewage flow that exceeds the capacity of the intercepting sewer will overflow to a surface waterway, perhaps a canal or river, or will flow into a combined sewer overflow conveyance or treatment facility before discharge to a waterway.

Sewersheds are not always by design, that is, they are not always based on a completely rational approach to the efficient layout of sewer networks for the collection and conveyance of sewage and stormwater. Municipal boundaries often dictate limits for a sewer system network because of ownership and maintenance responsibilities. As a city or village expands by annexation, sewers in the older part may flow one way while sewers in the newer section may flow another way. Sewer systems are usually built by land developers, and the sewer system is part of the planned development approved by a municipality. Topographic features, such as a ridge or higher ground surface elevations, may determine the location of the beginning of a sewer system. Branch sewers will be found on the periphery of a sewershed and larger trunk sewers will be found near the outlet.

Sewershed is not a commonly used term. Municipal officials and regulators at the federal and state level use the term *publically owned treatment works*, or *POTW*, to define all the sewerage infrastructure of a sewage treatment or water reclamation plant, including the sewer networks tributary to the plant. Another term frequently used is *facility area*, meaning the land area encompassed by a sewer network or system leading to the sewage treatment or water reclamation plant. The Chicago area includes several sewersheds such as the O'Brien plant sewershed, discussed next.

North Branch Watershed and the O'Brien Plant Sewershed

Farther upstream in the suburbs, away from the lakefront, the District dealt differently with the North Branch. As the villages of Glenview, Morton Grove, and Northbrook developed and grew in the early 1900s, residents sought relief as sewage discharging to the North Branch fouled the stream. In 1914, before the 1919 sewage treatment decision, the District built a small treatment plant in Morton Grove to abate pollution in the North Branch. This plant was located east of the river at the outlet of the village combined sewer, which was on the south side of Dempster Street. The initial technology from Germany, Imhoff tanks, combined settling and anaerobic solids digestion in one process. Later, in May 1920, an aerobic digestion process called a trickling filter was added. The plant included a settling tank to clarify the treated effluent and sand beds for drying the removed sewage solids, called sludge. Dried sludge was periodically scraped off the sand beds and used for fertilizer by local farmers.

Glenview and Northbrook sewer discharges were fouling the West Fork of the North Branch, so a treatment plant was built in Glenview; it was north of Henley Street along the West Fork. Placed in service in October 1924, the same technologies used in Morton Grove were employed. In Northbrook, construction of a treatment plant began in April 1924 on the east bank of the West Fork south of Walters Avenue, and it was placed in service one year later, again using the same technologies. The Glenview and Morton Grove plants served until July and December 1942, respectively, and the Northbrook plant was removed from service in January 1953. These plants were retired as the intercepting sewer system tributary to the O'Brien plant was extended as described above.

The North Branch natural watershed, best described by Libby Hill in *The Chicago River: A Natural and Unnatural History*, has its origin in Lake County near Park City. The watershed outlet is in the city at Wolf Point, just north of Lake Street where the North Branch flows into the Chicago River and/or South Branch. The outlet can be viewed from the Franklin or Lake Street Bridges or from the walkway along Wacker Drive between these two bridges. Once water gets into the North Branch or into one of its tributaries, that water will flow past the outlet. However, not all water in the watershed, whether from a hydrant or spigot or from precipitation, will flow past the outlet. Some water is diverted through sewers or tunnels to other sewersheds.

The entire North Branch watershed is long in the north-south direction, about 35 miles; it is narrow in the east-west direction, from two to five miles in Lake County and up to nine miles wide in Cook County. Geologically, the watershed lies between the Highland Park and Park Ridge moraines, which gradually disappear in the Chicago Lake Plain. The O'Brien plant sewershed boundary defines the watershed on the Chicago Lake Plain. At Touhy Avenue in Niles, upstream of the U.S. Geological Survey steam discharge measurement station, the watershed measures 100 square miles in area. Farther downstream, close to the North Branch Dam at the Albany Avenue station, the drainage area is 113 square miles. The additional 13 square miles account for storm sewers and riparian Forest Preserve property, which discharge directly to the river. However, most of what was the natural watershed downstream of Touhy Avenue is served by combined sewers, diverting stormwater through intercepting sewers to the O'Brien plant or via the Deep Tunnel to the Stickney plant.

In Lake County, although the character of the watershed has been significantly altered by extensive development, the watershed divide can usually be determined by following the highest land elevation on topographic maps. Sewers in Lake County are of the separate type; sanitary sewers are in one of two sewersheds conveying sewage to one of two water reclamation plants: the Village of Deerfield Water Reclamation Facility on Hackberry Road, which discharges to the West Fork of the North Branch about 0.5 mile north of Lake-Cook Road; and on Clavey Road in Highland Park, the long-named North Shore Water Reclamation District Clavey Road Water Reclamation Facility, which discharges to the Skokie River about 0.25 mile north of Lake-Cook Road. Sanitary sewers in the narrow strip of Lake Michigan watershed along the lakefront are part of the Clavey Road plant sewershed. Web sites for these two organizations can provide more detail.

Wet weather stormwater from developed areas throughout the watershed in Lake County, including major highways, reaches the North Branch, Skokie River, and West Fork via numerous separate storm sewersheds. Direct discharge from non-sewered areas is limited to forest preserves, golf courses, nature preserves, parks and undeveloped vacant land. However, stormwater originating in the narrow strip of Lake Michigan watershed along the lakefront drains to the lake. Most potable water used in homes and business comes from Lake Michigan via municipal water utilities or regulated private water utilities.

Cook County is different due to combined sewers and its Deep Tunnel, which, with one exception, begins south of Beckwith Road in Morton Grove, the first road crossing downstream of the confluence of the West Fork and the North Branch. The confluence of the Skokie River and North Branch is farther upstream near Happ Road. The small village of Golf, located north of Golf Road along the West Fork and with a population of about 500, is the only municipality in the watershed upstream of Beckwith Road that has combined sewers. Much like Lake County, the remainder of the watershed north of Beckwith is fitted with separate sanitary sewers and storm sewers. South of Beckwith Road combined sewers and storm sewers have completely altered the drainage patterns to conform to the grid. (See Figure 20.)

The area of the O'Brien plant sewershed is 143 square miles. The Cook-Lake County line on the north and the lakefront on the east are clear and easily identified boundaries of the sewered area. The south boundary is Fullerton Avenue from the lake to about Elston Avenue, and from 4600 west to Harlem Avenue. In between Elston and 4600 west, the south boundary is Bloomingdale Avenue. The west boundary is more erratic, but approximately described as follows: starting at Fullerton Avenue it is Harlem Avenue north to Foster Avenue, west to Oriole Avenue, north to Interstate 90, west to Canfield Avenue, north to Howard Street, east to Harlem Avenue, north to Monroe Street, west to Western Avenue, north to Dempster Street, west to Interstate 294, north to West Lake Avenue, west to Des Plaines River Road, north to Milwaukee Avenue, and north to the Cook-Lake County line. The boundary is known precisely, but liberties have been taken to simplify the description of the south and west boundaries.

The 143-square-mile sewershed is the gross service area and includes open areas such as cemeteries, golf courses, and forest preserves, as well as the combined and separate sewered areas. Some of the open areas are partially sewered. In dry weather, all sewage collected in separate sanitary and combined sewers flows to the O'Brien plant and, after the water is reclaimed, the water effluent is discharged to the canal just north of Howard Street. All the potable water used in the O'Brien sewershed comes from Lake Michigan; hence the effluent is reclaimed lake water with its own distinctive biological and chemical character similar to the effluent discharged from the two water reclamation plants in Lake County.

Separate sanitary sewer systems, when properly designed, constructed, operated, and maintained, will not be affected by rainfall. However, illegal connections of storm drains, building foundation drains, and/or roof downspouts can cause additional flow to surcharge the system, resulting in prohibited overflows to streets and/or sewer backups and basement flooding. The sewer system owner must inspect and maintain the sewers and enforce the prohibition on illegal connections. Failure to do so can result in sewer backup in basements or overflowing manholes in the street.

Dry weather conditions, which prevail most of the time, may be considered normal operation. Even with light rain, less than 0.1 inch, normal operation continues as in dry weather because little rainfall runoff occurs. Light rain wets the surface, clings to vegetation, and collects in small puddles. With more rainfall, excess rainfall runoff will enter combined sewers and will be accommodated within the intercepting sewers and at the O'Brien plant. Normal operation occurs about 90% of the time. Storm sewers in the separate sewered area receiving light rainfall may discharge to a waterway, and the added flow can be accommodated without flooding.

As rainfall continues, the flow in combined sewers increases and more of the sewage/stormwater mixture discharges at the control chamber to the intercepting sewer. When the intercepting sewer cannot accept more flow, the excess sewage/stormwater mixture will overflow to the North Branch or North Shore Channel. However, the combined sewer is connected to a Deep Tunnel drop shaft, and the tunnel gate will direct excess flow to the tunnel before overflow occurs. Excess flow to the Deep Tunnel will continue until the tunnel is full, then overflow will occur. While the intercepting sewer continues to convey combined flow to the O'Brien plant, the diverted excess flow to the Deep Tunnel bypasses the O'Brien plant and is eventually handled at the Stickney plant. Combined sewer overflow is discharged to the North Branch or North Shore Channel at numerous locations and will eventually exit the watershed in downtown Chicago.

If you live or work in a combined sewer municipality, your residential drainage finds its way to the combined sewer in a street via internal and underground pipes, while storm drainage finds its way to the same sewer via downspouts and surface flow to a curb inlet. Hopefully, your downspouts are not connected directly to the sewer. Normally, the street sewer flow ends up at the O'Brien plant and, after the water is reclaimed, it is discharged to the canal. On days with lots of rain, some of that street sewer flow will end up at the plant while the excess, which may be combined sewer overflow, is diverted to the Deep Tunnel or the North Branch or North Shore Channel. The dry weather route is known, but the wet weather route is usually indeterminate. See Table 3 to identify the type of sewer system for each municipality in the O'Brien plant sewershed.

If you live in a separate sewered municipality, your residential drainage ends up in the street separate sanitary sewer via internal and underground pipes and should always end up at the O'Brien plant, unless at some point it mixes with flow from a combined sewer and is diverted to the Deep Tunnel. On rainy days storm drainage from your home area ends up in the street storm sewer curb inlet via downspouts and surface flow. The storm sewer system will end up in a local surface waterway.

Municipal storm sewers are regulated by the Illinois Environmental Protection Agency through MS4 permits, the acronym for municipal separate storm sewer system. The permits do not contain numerical limits on biological, chemical, or physical constituents, however, a host of uses, practices, public education, monitoring, recordkeeping and reporting are required. The Cook County Department of Highways and Illinois Department of Transportation (IDOT) also own and operate storm sewers for efficient drainage of transportation corridors. Highway drainage is usually discharged to the nearest surface waterway or may be integrated with municipal storm sewers where the highway passes through the municipality. IDOT also owns and operates a unique storm drainage system for expressways. Coursing through the North Branch watershed are portions of Interstate Routes I-90 and I-94. Known as *main drains*, four exist within the watershed. They are long narrow sewersheds.

System number 1 drains the Edens Expressway (I-94) from Cherry Street in Winnetka to Dempster Street in Morton Grove, with a pumping station discharging to the Skokie River on the west side of the expressway south of Winnetka Road. System number two drains the remainder of the Edens from Dempster Street to Lawrence Avenue, with a pumping station beneath the expressway on Forest Glen Road discharging to the North Branch. System number five drains the Kennedy Expressway (I-94) from Canfield Avenue to Central Park Avenue, with a pumping station on Central Park discharging to the City's Roscoe Street relief sewer, which discharges to the North Branch at Roscoe Street. System number 9 drains the remainder of the Kennedy in the watershed, with a pumping station farther downstream outside the watershed.

Combined Sewer Stormwater Management and Deep Tunnel

Some refer to the District's Deep Tunnel as the Chicago Underflow Plan (CUP) or the Tunnel and Reservoir Plan (TARP). The acronyms have literal meanings, either providing a large receptacle or a protective cover. Despite the official sounding monikers, Deep Tunnel is the popular name. Planning for the Deep Tunnel began in the late 1960s by a consortium of state and local agencies under the District's leadership. The plan was approved locally in 1972 and received the first federal funding in 1974 under the Clean Water Act construction grants program; it was designed and constructed to serve only the combined sewer area.

Fundamentally, the Deep Tunnel is a supplement to the canal system to provide additional conveyance and storage capacity for combined sewer overflow, which is the mixture of sewage and stormwater that occurs with rainfall events that exceed the capacity of the intercepting sewers and treatment plants. Absent Deep Tunnel, combined sewer overflow would continue to pollute the surface waterways. Because of the additional conveyance and storage capacity, the Deep Tunnel also serves to reduce flood damages by providing additional outlet capacity for local combined sewers. It is designed, as required by federal and state environmental regulations, to remedy the pollution of the canal system and other waterways caused by combined sewer overflow.

Construction of the tunnels serving the O'Brien sewershed combined sewer area began in 1975 and was completed in 1998. The huge McCook Reservoir, still under construction in Hodgkins, Illinois, and designed to provide additional storage, is about 24 tunnel-miles southwest of the O'Brien plant. All combined sewer overflow captured by Mainstream tunnel system is pumped back to the Stickney plant and receives full treatment before being returned to the environment in the Sanitary & Ship Canal. One-third of the ten-billion-gallon storage capacity is due to be in service by 2017 and the remainder by 2029. (See Figure 21.) This reservoir serves two tunnel systems—Mainstream and Des Plaines. The total storage capacity of the reservoir not-yet-completed plus the already-completed tunnels is 12 billion gallons.

Twelve billion gallons is a volume difficult to comprehend! For comparison, it is equivalent to about 2.7 inches of water covering the contributing drainage area of 255 square miles. Another popular equivalent: 240 million 50-gallon rain barrels. If that number of rain barrels were lined up side by side, the line would stretch around the world more than three times. Locally, each tax parcel in the tributary area would need about 250 rain barrels. The area of the McCook Reservoir may be easier to comprehend—it is slightly smaller than Grant and Millennium Parks. Another equivalent: 17 stadiums the size of Soldier Field can be laid out on the reservoir floor.

The O'Brien plant sewershed is at the upper end of the Mainstream tunnel and is served by two tunnel segments. One is the Mainstream tunnel, which begins 22 feet in diameter 214 feet below ground at the Wilmette Pumping Station and follows the course of the canal to and beyond Fullerton Avenue to the Mainstream Pumping Station in Hodgkins, where the tunnel is 33 feet in diameter and 286 feet below ground. When the reservoir is in service, the tunnel will directly enter near the reservoir bottom. The second tunnel is a branch of the Mainstream tunnel and begins as 30 feet in diameter 202 feet below ground in the Linne Woods Forest Preserve in Morton Grove near the North Branch, south of Beckwith Road. The North Branch tunnel generally follows the course of its namesake river and joins the Mainstream tunnel south of Foster Avenue in the city, where it is 30 feet in diameter and 249 feet below ground.

At numerous locations along the course of these two tunnel segments, connections to the intercepting sewer and local combined sewers allow excess combined sewer flow to be diverted through vertical drop shafts to the Deep Tunnel. Each drop shaft is vented to the atmosphere and identified with signage. Vapor is often noticed rising out of the drop shaft grate at ground level. This is normal as the tunnel is humid, with a relatively constant temperature of about 55 degrees Fahrenheit. Most drop shafts are separated into air and water passages by vented vertical walls. The vents allow air to be drawn into the downward rushing water, trapping copious quantities of air, which helps to dissipate the tremendous amount of energy in the water when it hits the bottom of the shaft. The entrapped air is released and exits to the atmosphere. (See Figure 22.)

Without the McCook Reservoir in operation, during rainfall events of about one inch, the Mainstream and North Branch tunnel segments fill to capacity, but this varies due to antecedent conditions, location of the most intense rainfall, and the intensity of the rain event. The tunnel may be nearly full if there has been recent rainfall; or a heavy rain on the South Side may fill the tunnel even if it isn't raining on the North Side. Also, intense rainfall of more than one inch per hour fills the tunnel rapidly. In essence, without the reservoir available, the empty tunnel captures the first flush of combined sewer storm flow and, compared to pre-tunnel conditions, results in a reduction of the frequency and volume of combined sewer overflows. With reservoir storage available, the reduction will be much greater, likely eliminating most combined sewer overflows except for large intense rainfall events.

It is worth repeating that the Deep Tunnel and the associated reservoirs are solely for combined sewer areas. Also, the capacity of the tunnels and reservoir will not resolve all flooding problems in the O'Brien sewershed. The reservoir and tunnels may be partially full from previous storms when another storm rolls across the area, reducing the capacity to swallow the latest event. In addition, the local sewer infrastructure leading to the nearest tunnel drop shaft may have limitations to deliver excess flow in a timely manner.

Stormwater Management in Separate Sewered Areas

For many communities developed in the latter part of the twentieth century, housing was developed on flood-prone areas; that was before floodplain mapping identified these areas and regulations were put in place to restrict building. Dependence on automobiles brought on the need for wider streets and parking lots; that increased the amount of impervious surfaces on the landscape. Efficient drainage was required for these impervious surfaces, and as a result more stormwater gets to the rivers and streams, and it gets there faster. Urban flooding was and is commonplace in many areas, and despite the best efforts of the District, municipalities and other agencies to resolve flooding problems, flooding will continue when intense storms roll across the area.

Communities with separate sewers must rely on other means for the relief of flooding, and the District's stormwater management program meets this need. Drainage has been a fundamental responsibility of the District since its inception and it has always been looked upon to solve flooding problems. A flood control program was launched in the 1950s after the District's jurisdiction was doubled to serve fast growing suburban areas in Cook County. The plan, based on expediting the removal of floodwater, would have enlarged the size of channels and forced more floodwater into Lake Michigan. However, that would have been inconsistent with emerging concepts of managing stormwater near its source and not moving the flood problem "downstream." Furthermore, water pollution legislation at the federal level was leaning toward imposing stringent water quality standards, which made clear that combined sewer overflows would have to be abated, not increased.

In 1970 the District launched a new flood control program: it prepared watershed plans, placed limits on development in floodplains, restricted the discharge of stormwater from new developments, and implemented projects to capture and temporarily store floodwater in the separate sewered area of the District. The program was a companion to the Deep Tunnel for the already developed combined sewer area. More recent legislation in 2004 has specifically designated the District as the stormwater management agency for Cook County and has given it increased authority for this responsibility.

Based on the first round of watershed plans, the District constructed several floodwater detention reservoirs in the Cook County portion of the North Branch watershed:

- Glenview Reservoir, on the West Fork south of Willow Road
- Middle Fork Reservoir, on the North Branch, or Middle Fork, north of the Interstate 94 Edens Spur
- Northbrook Reservoir, on the West Fork north of Techny Road
- Techny Reservoir, on the West Fork north of Willow Road

The first three listed are excavated reservoirs with pumping stations; when streamflow rises above flood stage they fill by gravity and are pumped out following a storm event when the stream can safely accept the discharge. The Techny Reservoir is an inline reservoir behind a dam with a restricted outlet; rising water spreads laterally to occupy valley storage.

The Middle Fork Reservoir serves to relieve flooding problems in tributary areas in Northbrook and Northfield. The Glenview, Northbrook, and Techny Reservoirs, in series on the West Fork, serve to relieve flooding problems in tributary areas in Glenview, Golf, Morton Grove, and Northbrook. In addition to these large projects, the District has cooperated with municipalities in building smaller reservoirs.

A new North Branch watershed management plan, prepared in 2008, identifies additional flood control projects for implementation. In addition to planning and construction, the District administers an updated watershed management ordinance, funds flood damage reduction and best management or green infrastructure projects initiated by municipalities, and purchases flood prone property that cannot cost-effectively be protected by flood control improvements. Critical to the efficacy of improvements is a program known as the District's small streams maintenance program; it maintains existing waterways by working in cooperation with local municipalities to remove debris blockages in local waterways. The District's web site, mwrd.org, includes an explanation of its stormwater management program.

Stormwater management in the Lake County portion of the North Branch watershed is handled by the Lake County Stormwater Management Commission. Under state authority, the commission provides services similar to the District. More information can be found on their web site.

Who Pays for What

Throughout the water utility industry, rate payers are the primary sources of revenue for administration, capital improvement, maintenance, and operation. Periodically the utility sends the resident or business a statement that reflects the quantity of water consumed, reports an impervious surface measure or other unit measure, then applies a rate to calculate the amount owed for water supplied and for sewerage service and/or stormwater service. Thus, the amount owed is related to the amount used and to the amount of impervious area. Needless to say, this is an incentive to conserve. However, utilities have large fixed costs; if conservation lowers usage payments, then rates may be driven upwards.

The District is different. Alone among other large utilities, it relies on property taxes for over 80% of its revenue. The cost for District services is based on the value of real property owned, not on consumption. There is merit in this method as the District does not have to support the significant cost of billing and collection; revenue is collected by the Cook County Treasurer. Landowners with expensive property may not use more water or cause more stormwater runoff, but they pay more because of the assessed value of their ownership. This too has merit because some cost is shifted to the more affluent, relieving the burden on the less affluent. One argument in support of the method is that the value of District service adds to the value of individual ownership as well as to the overall economic wellbeing in the District's service area.

There is somewhat of a tug-of-war among the city, the suburbs (especially the separately-sewered ones), and the suburbs with mostly combined sewers over who pays for what. Some in the suburbs object to paying for the expensive deep tunnels and big reservoirs that only benefit the city. However, the District has flexed its muscle in Washington to receive, through federal funding and low-interest loans, about half of the cost of the tunnels and reservoirs. On the other side, some in the city object to paying for the suburban stormwater channel improvements, reservoirs, and small streams program. A characterization, certainly apt for this topic, is that one hand washes the other; the entire service area benefits by what the District does and how they do it.

The north area as used herein refers to the area of the District in the O'Brien plant sewershed and is distinctly different than the North Service Area as used by the District. The latter includes three other water reclamation plants spread across north Cook County and will be discussed in another book.

References

Chicago Tribune Archives. archives.chicagotribune.com.

deefield.il.us.

Evanston History Center. Building permit files and city annexation map.

lakecountyil.gov/stormwater.

Metropolitan Sanitary District of Greater Chicago. *The Story of the Metropolitan Sanitary District of Greater Chicago: The Seventh Wonder of American Engineering*, 1959 Edition.

Metropolitan Water Reclamation District of Greater Chicago. Engineering Department Sewer Atlas, 2012.

Metropolitan Water Reclamation District of Greater Chicago. Maintenance and Operations Department Annual Report, 1983 and 1984.

Metropolitan Water Reclamation District of Greater Chicago. Maintenance and Operations Department Facilities Handbook, 2012.

Metropolitan Water Reclamation District of Greater Chicago. Proceedings of the Board of Commissioner/Trustees, 1890 through 2015.

mwrd.org.

northshorewrd.org.

Sanitary District of Chicago. *Engineering Works*. 1928.

wikipedia.org.

Table 3: Municipal and Township Local Sewer Systems

Municipality	Sewer System
Chicago	Combined
Evanston	Combined
Glencoe	Separate
Glenview	Separate
Golf	Combined
Kenilworth	Both
Morton Grove	Both
Niles	Both
Northbrook	Separate
Northfield	Separate
Skokie	Combined
Wilmette	Both
Winnetka	Separate

Township	Sewer System
Evanston	Combined
New Trier	Separate
Niles	Separate
Northfield	Separate
Wheeling	Separate

Notes:

The township sewer system is shown for unincorporated areas. Local sewer systems may be under the jurisdiction of a municipality, township, county, sanitary district, or private utility.

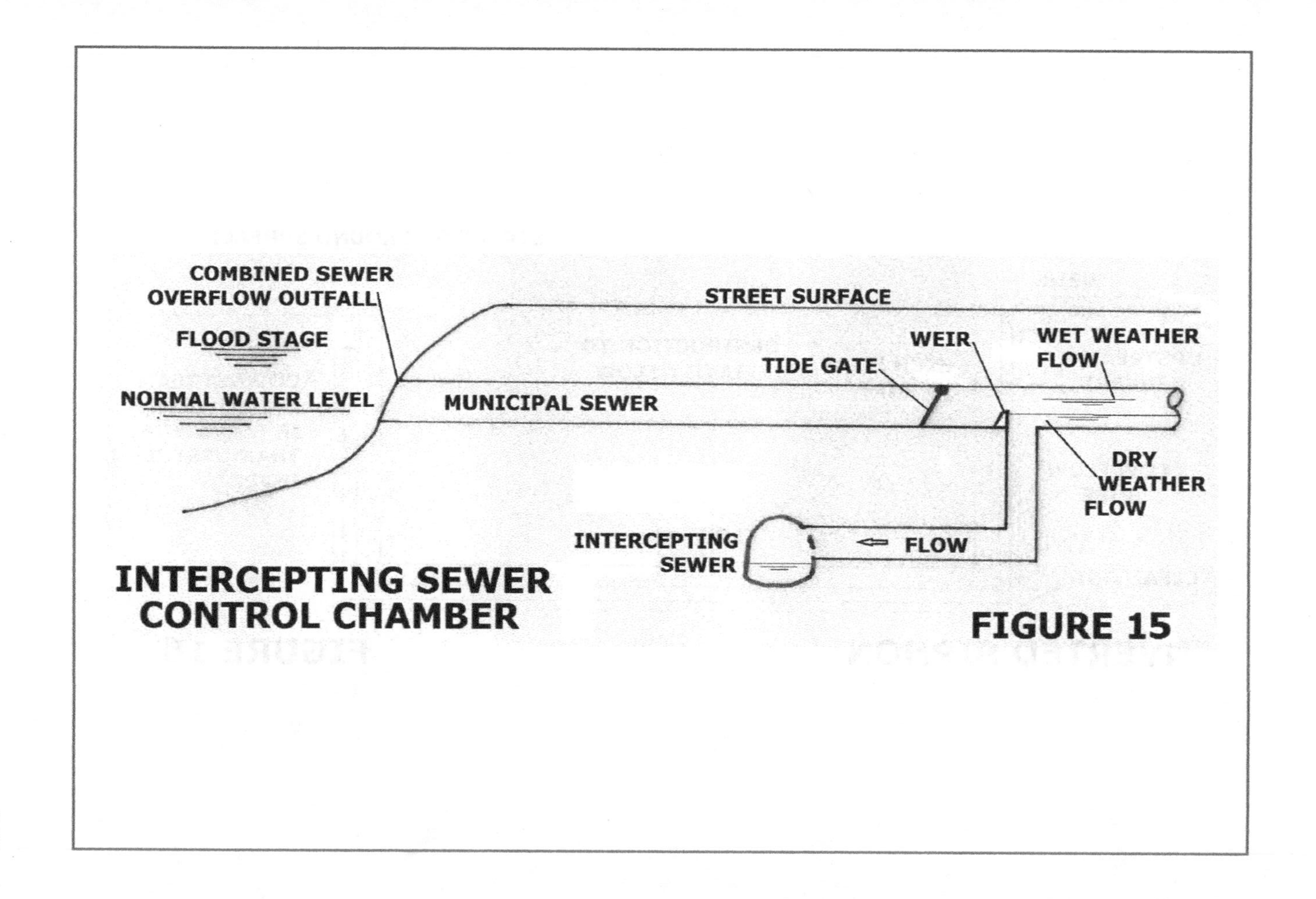

INTERCEPTING SEWER CONTROL CHAMBER

FIGURE 15

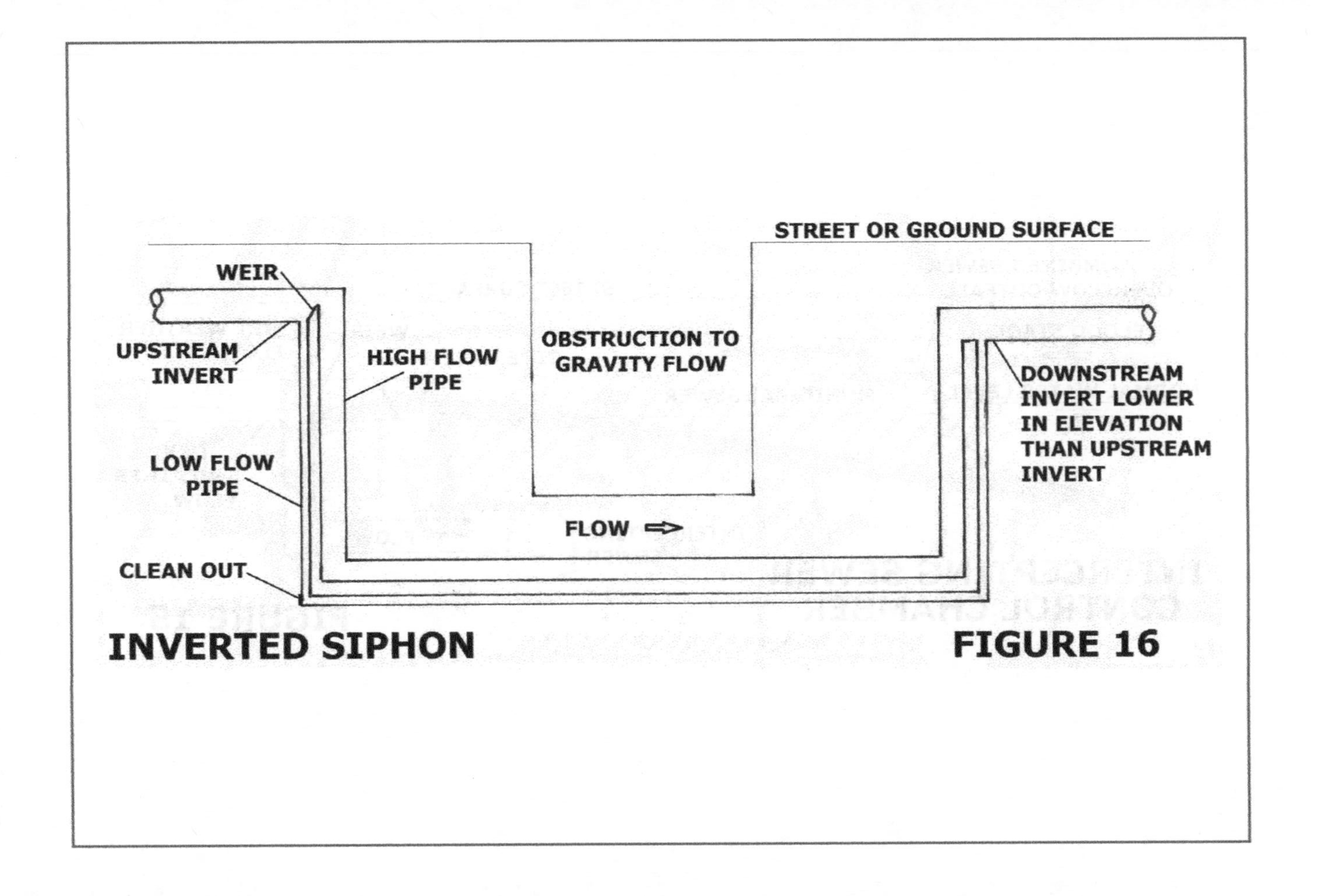

INVERTED SIPHON FIGURE 16

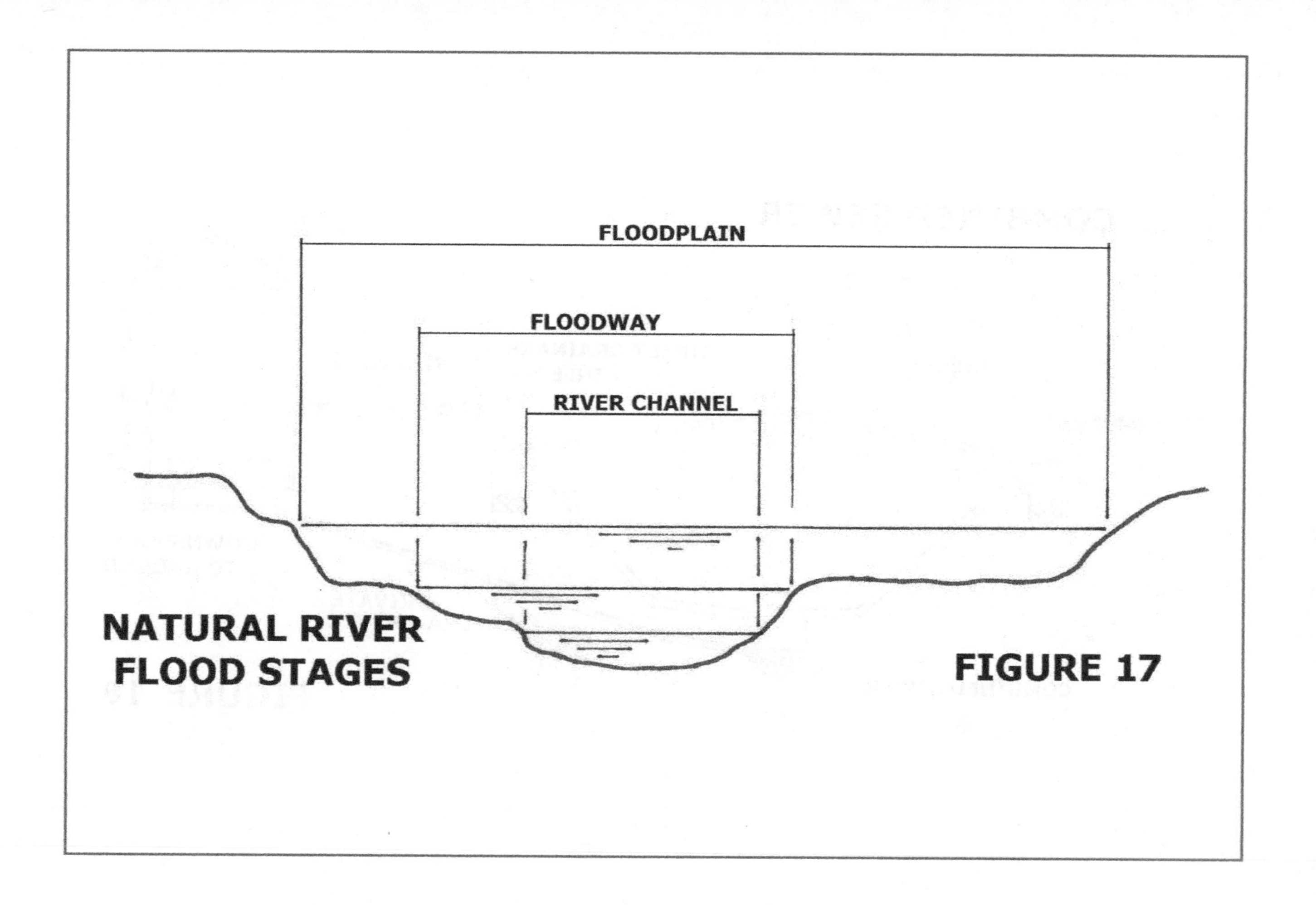

NATURAL RIVER FLOOD STAGES

FIGURE 17

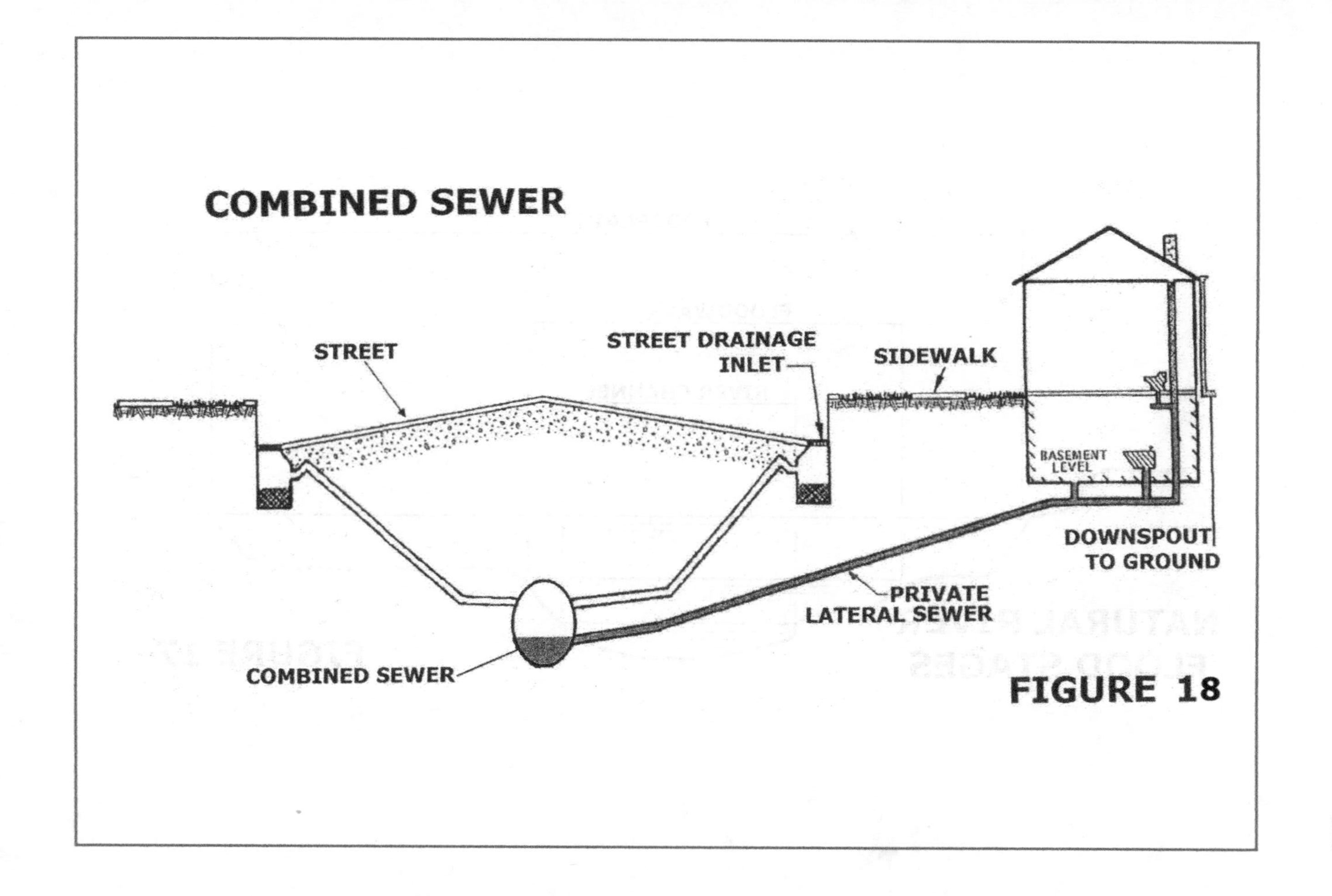
COMBINED SEWER
STREET
STREET DRAINAGE INLET
SIDEWALK
BASEMENT LEVEL
DOWNSPOUT TO GROUND
PRIVATE LATERAL SEWER
COMBINED SEWER

FIGURE 18

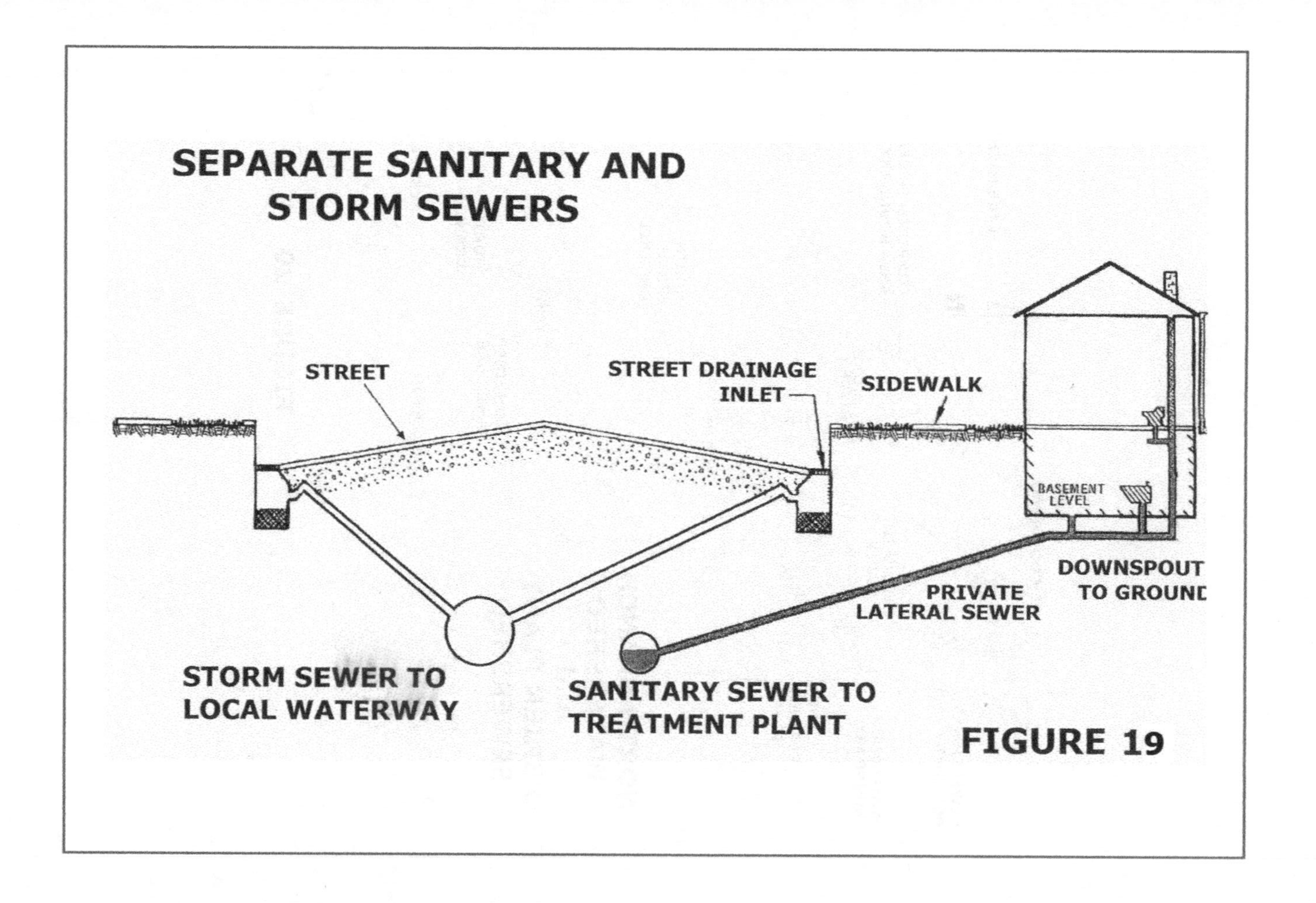
SEPARATE SANITARY AND
STORM SEWERS
STREET
STREET DRAINAGE
INLET
SIDEWALK
BASEMENT
LEVEL
DOWNSPOUT
TO GROUN
PRIVATE
LATERAL SEWER
STORM SEWER TO
LOCAL WATERWAY
SANITARY SEWER TO
TREATMENT PLANT

FIGURE 19

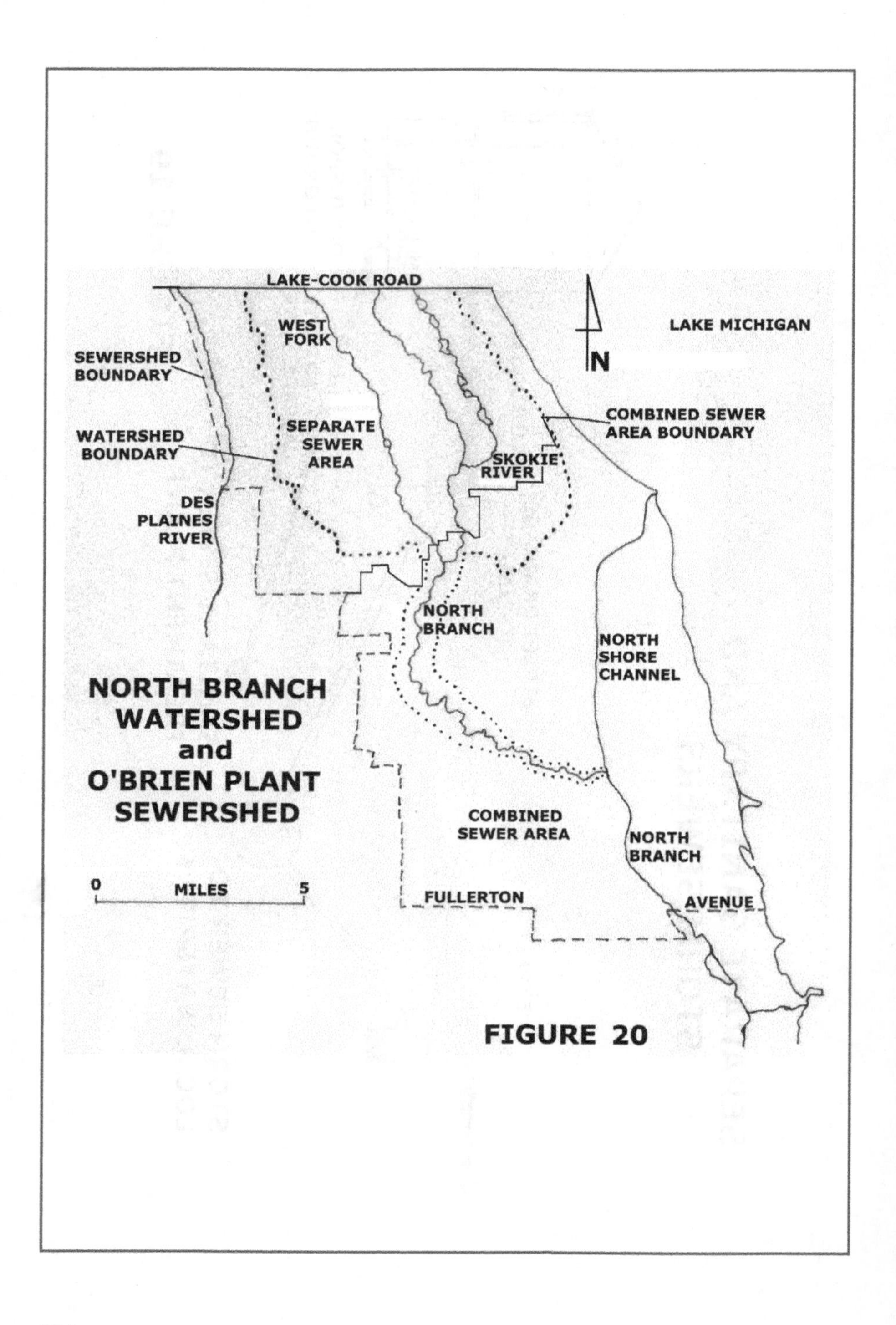
LAKE-COOK ROAD
LAKE MICHIGAN
N
WEST FORK
SEWERSHED BOUNDARY
WATERSHED BOUNDARY
SEPARATE SEWER AREA
COMBINED SEWER AREA BOUNDARY
SKOKIE RIVER
DES PLAINES RIVER
NORTH BRANCH
NORTH SHORE CHANNEL
NORTH BRANCH WATERSHED and O'BRIEN PLANT SEWERSHED
COMBINED SEWER AREA
NORTH BRANCH
0 MILES 5
FULLERTON
AVENUE

FIGURE 20

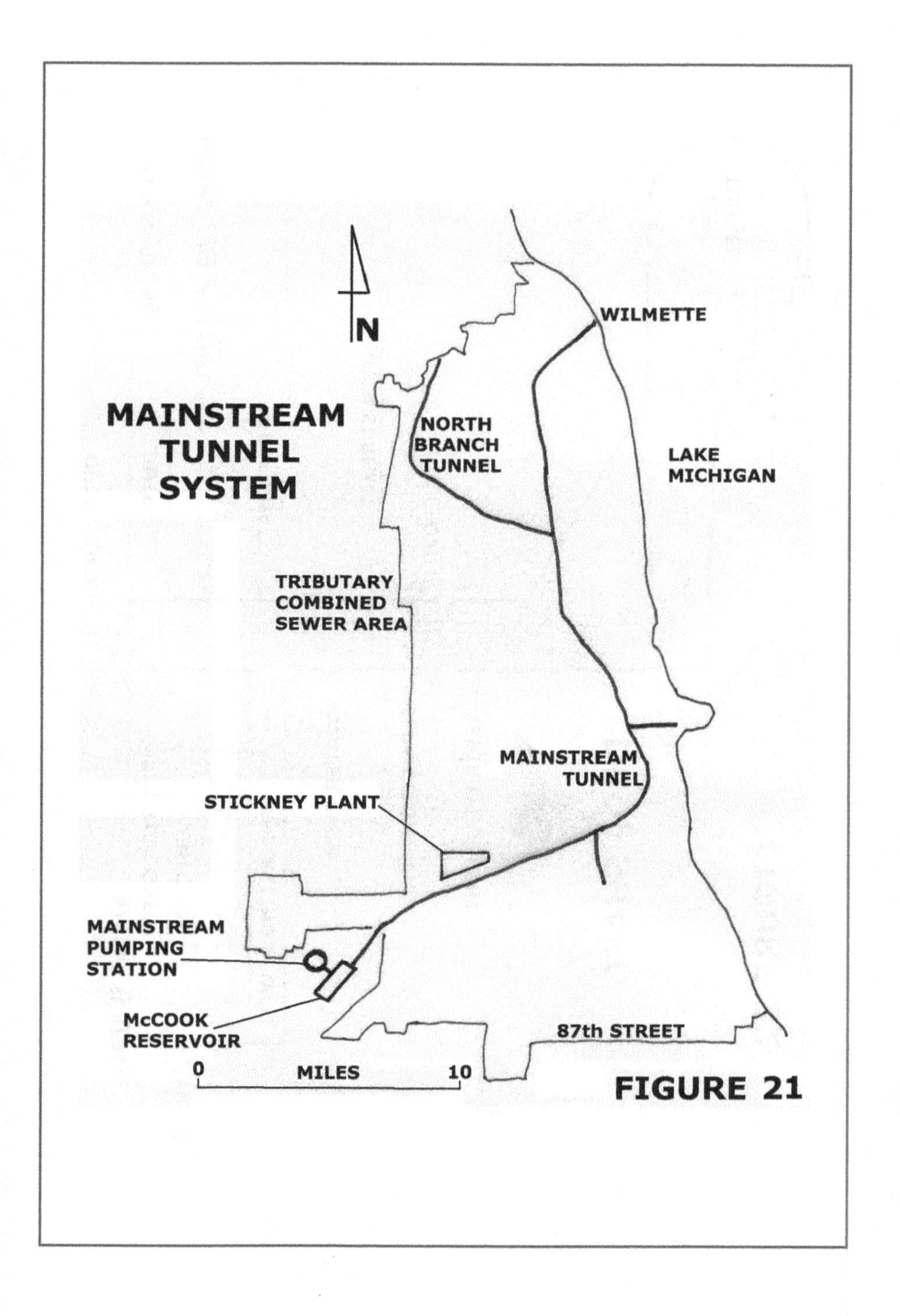
N
WILMETTE
MAINSTREAM
TUNNEL
SYSTEM
NORTH
BRANCH
TUNNEL
LAKE
MICHIGAN
TRIBUTARY
COMBINED
SEWER AREA
MAINSTREAM
TUNNEL
STICKNEY PLANT
MAINSTREAM
PUMPING
STATION
McCOOK
RESERVOIR
87th STREET
0
MILES
10

FIGURE 21

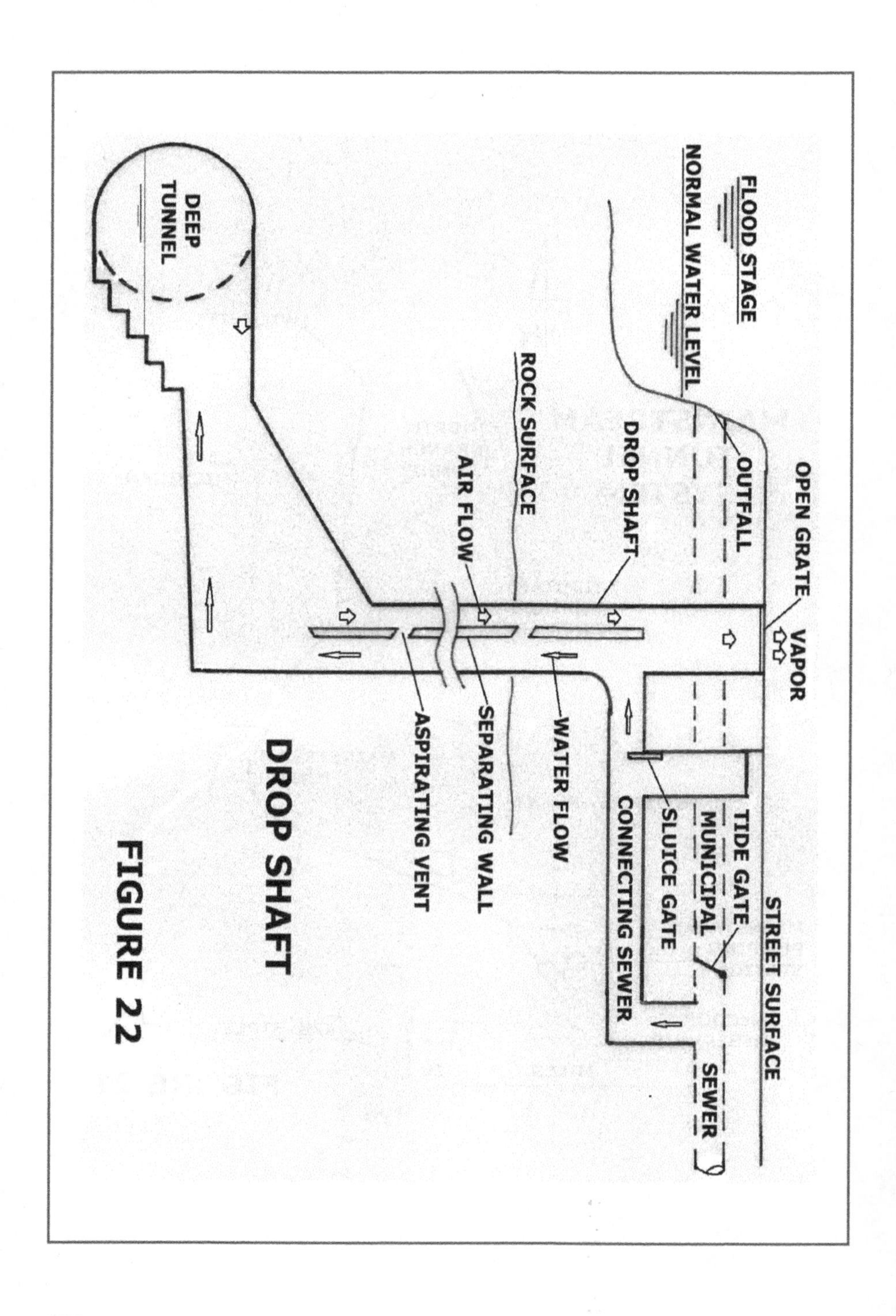

DROP SHAFT

FIGURE 22

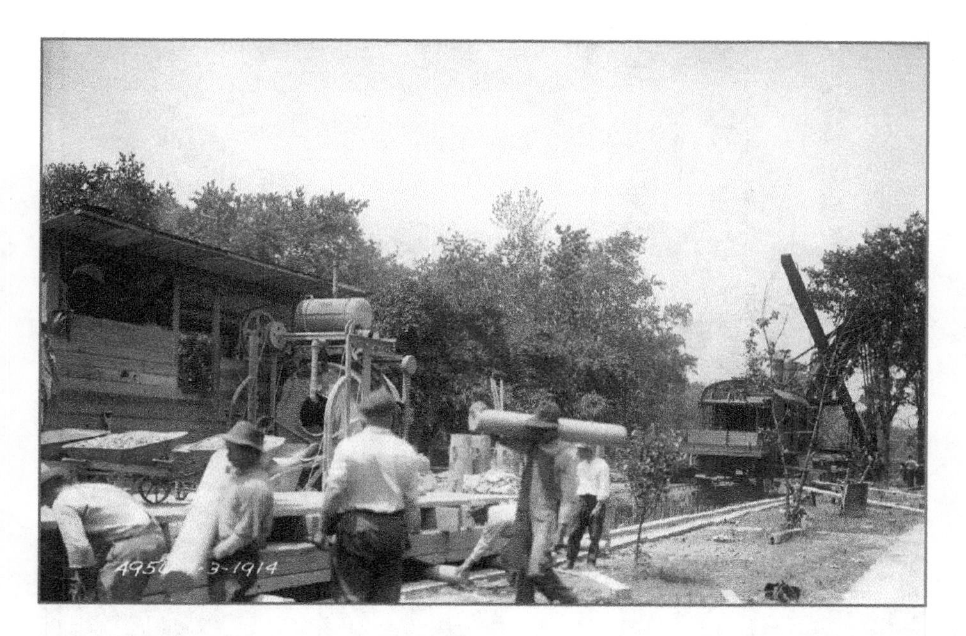

Photograph 4.1: Constructing the North Shore Intercepting Sewer by open cut along Sheridan Road in Wilmette on July 3, 1914. A steam shovel at right is excavating the trench, and the concrete mixer at left follows. The skid-mounted mixer is moved along using the rollers carried by the laborers. (MWRD photo 4956)

Photograph 4.2: The soil on the sides of the trench are held in place by the steel sheeting and horizontal bracing while a movable form is moved by the trolley on July 3, 1914, for the intercepting sewer in Wilmette. The bottom segment of the sewer cross-section is cast first followed by the upper segment shown in the background. (MWRD photo 4957)

Photograph 4.3: George Wisner, District chief engineer, addresses a crowd for the groundbreaking of the Evanston Intercepting Sewer at 3:00 pm on August 23, 1916, at the intersection of Custer and Mulford Streets in Evanston. Several trustees and local officials were on the podium and preceded Wisner in speaking. A local band played music and refreshments were served in the tent. (MWRD photo 5606)

Photograph 4.4: A District engineer inspects the designated area for spoil disposal for the intercepting sewer construction in Evanston along the lake shore between Church and Davis Streets on January 2, 1917. The church steeple and buildings to the north in the background are part of the campus at Northwestern University. (MWRD photo 5744)

Photograph 4.5: Carpenters are building forms for the reinforced concrete foundation of the Evanston Pumping Station on October 23, 1919. A stiffleg derrick is used to lower materials and a concrete mixer stands ready next to the field office. The house on the right across Elmwood Avenue no longer exists; the site is presently occupied by the Evanston Police Department headquarters. Some of the houses south of Lake Street remain to the present time. (MWRD photo 7322)

Photograph 4.6: The completed Evanston Sewer Pumping Station viewed on May 5, 1921, from the southwest. It is an important public works asset and a handsome addition to architectural styles in Evanston, with Prairie School features in red brick with limestone detailing. Sewers east of Chicago Avenue in Evanston are too low in elevation for gravity flow to the North Shore Channel, hence the need for a pumping station. (MWRD photo 8163)

Photograph 4.7: May 31, 1922. Most of the deep intercepting sewers were constructed by tunnel because this method was less costly than the open cut method. Deep intercepting sewers were necessary to receive sewage from local municipal sewers. The tunnel is accessed through vertical shafts at intervals along the tunnel length. Tunnel headings were mined in each direction from the shafts meeting the headings from the adjoining shafts. This head house on North Side Intercepting Sewer No. 1 covers the top of a vertical shaft near Isabella Street in Evanston. A steam hoist is used to move laborers, equipment and spoil in the shaft. The spoil in small dump cars is raised to the surface and then towed down the horizontal track and dumped, later to be hauled away, perhaps to the lakefront for disposal. (MWRD photo 9117)

Photograph 4.8: Laborers at the face of the tunnel heading are mining clay one block north of Main Street on August 30, 1922, for North Side Intercepting Sewer No.1. The soft moist clay could be carved like butter. The slabs of clay were placed in dump cars, and the cars moved to the shaft and hoisted to the surface. Timber bracing and wood sheeting were installed as mining progressed to prevent the tunnel from collapsing. (MWRD photo 9427)

Photograph 4.9: August 30, 1922. After mining out the clay, forms were built in the bottom of the tunnel, and the concrete invert of the intercepting sewer was poured. When the invert was cured, steel braces in the form of an arch were set, wooden planks attached, and concrete poured behind the wooden planks at the north heading, a block north of Dempster Street. (MWRD photo 9423)

Photograph 4.10: A finished section of intercepting sewer tunnel looking north on August 30, 1922. The interior bracing and forms have been removed, and the concrete reveals the outline of the form boards. The North Side Intercepting Sewer No.1 extended from south of Oakton Street north to Sheridan Road in Wilmette. It was 6.5 feet high south of Lake Street and 4.5 feet high north of Lake Street. (MWRD photo 9424)

Photograph 4.11: The intercepting sewer under construction through Evanston passes under West Railroad Avenue approximately where the fire hydrant is located. Tunneling invariably resulted in some surface subsidence, shown by the slight dip in the sidewalk. The Chicago & North Western Railroad is to the right in this north facing view on April 7, 1922. The houses along the west side of the street have been replaced by Kingsley Elementary School. (MWRD photo 8848)

Photograph 4.12: North Side Intercepting Sewer Nos. 3 and 4 are the largest size, 15 feet high, running from Howard Street south to Argyle Street. Laborers are mining clay on the north heading midway between Howard Street and Touhy Avenue, on September 20, 1924, and the size requires two levels of work. A loaded dump car is full and ready to be pushed south to the shaft for hoisting to the surface for disposal. Concrete placement keeps pace with the mining. (MWRD photo 11190)

Photograph 4.13: Looking at the south heading in the intercepting sewer south of Touhy Avenue on February 14, 1925, the bulkhead has failed and a few of the lattice arch supports have buckled. The ground above also subsided. The contractor will excavate the unstable soil from above, replace the forms and supports, and continue mining the heading and pouring the reinforced concrete sewer. After the concrete sewer has cured, backfill will be placed in the hole from above and the ground surface restored. (MWRD photo 11529)

Photograph 4.14: The contractor has completed the north end of North Side Intercepting Sewer No. 3 at Howard Street on April 1, 1925, and has installed a bulkhead to seal the end of the section of sewer. The bulkhead will remain in place for two or three years until another contractor completes the connection to the influent screen chamber for the North Side Sewage Treatment Works. (MWRD photo 11651)

Photograph 4.15: September 6, 1924. The bottom of a typical construction shaft is within the bottom portion of the completed sewer. Two hoists are in active use to maintain the supply of construction materials and removal of mined spoil. The center vertical strut separates the two hoists, and the hoisting engines are topside in the shaft head house. (MWRD photo 11167)

Photograph 4.16: The many construction shafts must be sealed before a contract is complete. All hoisting machinery has been removed from this shaft on April 14, 1925, and the contractor will install forms and pour the reinforced concrete crown of the sewer. When cured, the shaft above will be backfilled and the surface restored. (MWRD photo 11655)

Photograph 4.17: August 20, 1925. East of the intersection of Albany and Carmen Avenues, North Side Intercepting Sewer No. 3 crosses under the North Branch, upstream of the North Branch Dam. The sewer in this short segment was constructed by the open cut method to avoid problems with water and the potential for flooding. The stream was conveyed over the sewer construction in a wooden flume shown here to the south behind the horizontal bracing. Parts of the North Branch Dam are visible to the left and the completed sewer is in the foreground. (MWRD photo 12088)

Photograph 4.18: Looking west on August 20, 1925, the flume is in the center behind the industrial track used to transport material and spoil. Open cut sheeting follows the sewer alignment on both sides of the flume. Beyond the flume, standing water in the North Branch is caused by construction debris in the foreground obstructing the flow. (MWRD photo 12089)

Photograph 4.19: Looking southeast across District property and the right-of-way of the intercepting sewer on June 27, 1929, in the Ravenswood Manor neighborhood. Manor Avenue is on the right approaching the intersection with Giddings Street. This lot and the lot across the alley to the north are perhaps the only private properties acquired for the North Side Intercepting Sewer. For years, the District would only cut the weeds when neighbors complained. In recent years, the lot is leased and maintained as a neighborhood park by the Ravenswood Manor Improvement Association, and has been named LaPointe Park. (MWRD photo 15760)

Photograph 4.20: North Side Intercepting Sewer No.6 from Irving Park to Barry Avenue crossed the intake and discharge conduits for cooling water at the Commonwealth Edison Northwest Generating Station in California Avenue south of Addison Street. The Edison conduits were stacked vertically and the upper conduit was directly in line with the sewer. Edison rebuilt their conduits at District expense so the upper conduit was next to the lower conduit, and the District constructed the sewer atop the conduits. Looking east on October 21, 1925, the two conduits are exposed, the deep excavation is suitably braced, and forms for the new conduit are under construction. (MWRD photo 12213)

Photograph 4.21: Looking north on December 19, 1925. The new conduit has been completed, and Edison added sluice gates in each to facilitate maintenance. The District proceeded to build the sewer crossing over the conduits within the right-of-way of California Avenue. The conduits were used for cooling water from the North Branch until 1970 when the generating station was taken out of service. (MWRD photo 12464)

Photograph 4.22: Looking north showing the heading and completed sewer in North Side Intercepting Sewer No.6 under California Avenue about midway between Melrose Street and Otto Street on April 13, 1925. A District construction engineer is inspecting the contractors' work in this 10.5-foot-high sewer. (MWRD photo 11664)

Photograph 4.23: January 30, 1925. A view of North Side Intercepting Sewer No. 6, looking east under Melrose Street east of California Avenue, shows the completed 10.5-foot-high sewer forms, where concrete was recently poured, and the heading with exposed clay. The sewer was routed via Melrose Street and Washtenaw Avenue to avoid the complex Belmont/California/Elston intersection. (MWRD photo 11476)

Photograph 4.24: North Side Intercepting Sewer No. 6 looking northwest around the curve under the intersection of Melrose Street and Washtenaw Avenue on April 13, 1925. This section is completed except for the floor, which has been filled in for a flat surface while construction continues. The false floor will be removed before contract completion. (MWRD photo 11661)

Photograph 4.25: Looking south at the Belmont Avenue junction chamber under construction on April 13, 1925. The intercepting sewer reduces from 10.5 to 8.5 feet in height on the right, as the flow from the Belmont Avenue sewer enters from the left. This junction chamber is in Washtenaw Avenue just north of Belmont Avenue. (MWRD photo 11662)

Photograph 4.26: The outfall for the Belmont Avenue City sewer discharges to the North Branch at the southwest corner of the bridge. Sewage in this sewer is intercepted at Washtenaw Avenue. Looking west across the channel on February 19, 1925, the outfall is at the stern of a docked boat. The Concordia Evangelical Lutheran Church in the background was built is 1892. (MWRD photo 11542)

Photograph 4.27: Looking east in George Street from Elston Avenue on November 3, 1925. A shaft house is under construction for North Side Intercepting Sewer No. 7. The intercepting sewer was 7.5 feet high from Barry Avenue to the Diversey/Elston/Western intersection and five foot high thereafter to Fullerton Avenue. (MWRD photo 12275)

Photograph. 4.28: This shaft is located at Rockwell Street and Barry Avenue on North Side Intercepting Sewer No.8. Looking northwest, the contractor is beginning to mix a load of concrete for sewer construction on March 18, 1926. Three segments of No. 8 were built in various sizes on the east side of the North Branch and connected through inverted siphons to the intercepting sewer on the west side of the North Branch. The building in the background is one of several small tanneries in the area. (MWRD photo 12535)

Photograph 4.29: Perhaps 8,000 of these standard sewer manhole covers are spread throughout Cook County, controlling access to the underground network of intercepting sewers, control chambers, and other vaults. Weighing about 70 pounds, the weight is only one deterrent to removing the cover, as each cover is locked with a simple yet clever mechanism. This cover is between 26 and 60 years of age as revealed by the name. The 21-inch-diameter opening in the frame into which the cover rests limits the size of people and equipment going underground. (Photo by the author)

Photograph 4.30: Construction of the Morton Grove Sewage Treatment Works looking southeast from the Dempster Street Bridge over the North Branch on November 14, 1919. The original plant consisting of an Imhoff tank, small building, and sludge drying bed in the center, is being expanded to the right with a trickling filter in the building and final settling tanks. The sludge drying bed is also being expanded in the center. (MWRD photo 7374)

Photograph 4.31: By September 12, 1920, the Morton Grove plant was completed and the discharge to the North Branch obscured by the overgrowth. The Morton Grove station on the Chicago, Milwaukee & St. Paul Railroad, presently the Metra Milwaukee District North Line, is a block south along the tracks at right. (MWRD photo 7820)

Photograph 4.32: The Glenview Sewage Treatment Works has been in service three years on October 24, 1927. In the left foreground is the covered Imhoff tank and the open sludge drying beds. At center is the pumping station building and covered final settling tank. The open trickling filter with sprayers in operation is in the right background. (MWRD photo 13533)

Photograph 4.33: The Northbrook Sewage Treatment Works are under construction on March 26, 1925, nearly a replicate of the Glenview plant. All three of these suburban plants were removed from service when the intercepting sewer network reached the municipality and sewage could flow by gravity to the plant in Skokie. (MWRD photo 11642)

Photograph 4.34: Construction of the North Side Sewage Treatment Works began in 1923, and the first work was laying a 24-inch drain for the site north of Howard Street and west of the North Shore Channel. A steam shovel excavating the drain trench is loading spoil into dump cars on September 27, 1923. The drain discharged to the channel. (MWRD photo 10114)

Photograph 4.35: Excavation for the huge aeration battery C at the North Side Sewage Treatment Works was well underway on October 19, 1923, when a storm caused the excavation to fill with floodwater, inundating a steam shovel. To the north in the background is a gas holding tank, smokestack, and building at the manufactured gas plant on Oakton Street. (MWRD photo 10204)

Photograph 4.36: The contractor's construction yard on July 17, 1924. Looking northeast from the top of the south cableway tower shows the spur line off the Mayfair Division railroad to the right and additional spurs flanked by materials for construction of the plant. At left is the manufactured gas plant and in the center background across the North Shore Channel is a brick making plant, both being on Oakton Street. Also, from left to center is construction of the embankment for a new electric line, presently the Chicago Transit Authority Yellow Line, to the point where a bridge will be built over the channel. (MWRD photo 10931)

Photograph 4.37: Beginning the pouring of concrete for the deep foundation of the Pump and Blower Building on April 8, 1926. The contractor is using guyed tower cranes, the type typical of the time. The taller more distant tower is adjacent to the concrete mixing plant and is used to aerially transfer batches of concrete to the shorter closer tower, from which the concrete flows via the long sloping pipe. The outstanding arm of the shorter tower is also used for hoisting materials. (MWRD photo 12548)

Photograph 4.38: The Pump and Blower Building pump suction chamber construction, looking northwest on April 22, 1926, about 40 feet below grade. The lineup of reinforcing bars indicates where walls in the pump suction chamber will be built to guide the influent sewage to pump intakes, and the circular formwork shows the location of one of six cast iron pump suction bells. (MWRD photo 12606)

Photograph 4.39: Erection of the structural steel skeleton of the Pump and Blower Building for the plant, looking northwest on October 29, 1926. The skeleton will be enclosed with brick masonry and limestone detailing. The pumps and blowers are in a large sky-lighted interior atrium. (MWRD photo 13092)

Photograph 4.40: December 27, 1926. The structural skeleton has been completed, but the masonry walls must wait for warmer spring weather. A rail spur into the building at right on the blower floor facilitates equipment deliveries and unloading with the building gantry crane. On the lower pump floor at left, the intake and discharge piping awaits installation of the pumps and motors. (MWRD photo 13126)

Photograph 4.41: August 2, 1928. In this overview of the Pump and Blower Building interior, only four of the six pumps and five of the seven blowers are completely in view. The sixth blower in the right foreground is not completely assembled. The seventh blower and two pumps are out of view. (MWRD photo 14207)

Photograph 4.42: The Pump and Blower Building exterior, and what appears to be a main entrance, are viewed from the southeast along McCormick Road on October 16, 1927. The decorative entrance is not used; the working entrances are on the north and west sides of the building. Howard Street, an unimproved two-lane road at this time, is just out of view on the left. As shown by the sign, John Griffith & Son Company, Builders, was the contractor for this building. (MWRD photo 13506)

Photograph 4.43: The exterior wall of the southeast corner of the Pump and Blower Building is still being installed on February 3, 1927, while interior electrical and mechanical work is underway using the large opening on the west side of the building for access. The rail spur in the building shown in Photograph 4.40 was routed through this opening. The construction activity at the southwest corner is for the underground sewage discharge conduit that runs under the Mayfair Division tracks to the west. Air from the blowers exits the building underground at the northwest corner of the building. (MWRD photo 13177)

Photograph 4.44: The sewage conduit from the Pump and Blower Building runs west underground to the Grit Building, where it diverges into several channels for the grit tanks and fine screens. On July 20, 1926, the floor of the diverging structure has been completed and ironworkers are preparing the reinforcing rods for wood forms for the exterior confining walls and internal guide walls. (MWRD photo 12858)

Photograph 4.45: The Grit Building viewed from the east on January 27, 1927, is being enclosed with masonry walls and a flat roof with skylight. The outline of the diverging structure can be seen immediately east of the building, and it will be covered with a sidewalk and road. The preliminary settling tanks are barely visible west of the building; the aeration batteries are at right. At right, the cableway tower is used for work on the preliminary tanks. (MWRD photo 13176)

Photograph 4.46: Looking east at the discharge side of the Grit Building on August 18, 1926. Channels from grit tanks and three fine screens converge to flow on to the preliminary settling tanks. Each channel is controlled by a sluice gate on the intake and discharge sides of the Grit Building so individual tanks can be taken out of service for maintenance. (MWRD photo 12913)

Photograph 4.47: Eight square preliminary settling tanks were originally constructed, and seven are shown in this view on October 16, 1927. Square tanks were preferred because they were an efficient use of land. However, circular settling tanks have proved to be more operationally efficient; thus, eight circular preliminary settling tanks were added in 1988 west of the square tanks. The Main Building in the center houses the pumps for transmitting sludge to the Stickney Plant as well as other operating equipment for the activated sludge process. Administrative offices and a laboratory were located in the tower. Offices were relocated into a new building in 1999 and the laboratory moved to another plant. (MWRD photo 13057.1)

Photograph 4.48: An empty preliminary settling tank on June 27, 1928, shows the raking mechanism. As the arms slowly rotate, the blades on the bottom move the settled sludge toward the center of the tank, where it drops into a sump for removal. Preliminary and final settling tanks operate similarly, but the composition of sludge is different. (MWRD photo 14105)

Photograph 4.49: The preliminary settling tanks are filled with water on August 2, 1928, and ready for operation in two months. The bridge crossing each tank secures the upper end of the vertical axis for the underwater raking mechanism, and the house on the bridge contains the motor and gear reduction unit for rotating the rake. (MWRD photo 14210)

Photograph 4.50: Originally, the twin air mains ran underground from the Pump and Blower Building to the aeration tanks. A worker is using a surveyor's level to set the 60-inch casti-iron air mains at the proper elevation on November 22, 1923, near the aeration tanks. The connection to the blowers was not made until 1927. (MWRD photo 10324)

Photograph 4.51: A cableway was set up in the north-south direction to deliver equipment and materials for construction of the aeration tanks, operating gallery, and final settling tanks. Here on June 10, 1924, looking south, a five-cubic-yard concrete bucket assists in pouring the floor of the aeration tanks for battery C. (MWRD photo 10698)

Photograph 4.52: The south end section of two aeration tank sidewalls in battery C have been completed, two ironworkers are preparing the reinforcing rods for the next section of sidewall, and other ironworkers are preparing the south end wall reinforcing rods and wooden forms for concrete placement on August 5, 1924. (MWRD photo 11034)

Photograph 4.53: The prefabricated steel forms for the 15-foot high aeration tank walls were reusable. On July 22, 1924, at an intermediate wall in battery C, ironworkers are making final connections before concrete will be poured for this section of the wall. (MWRD photo 10959)

Photograph 4.54: The south cableway tower stands tall above the plant site to the south of battery C, where aeration tank walls are under construction on July 22, 1924. The tower rides on the two straight tracks to the left. The photographer is elevated on the top of the concrete mixer, from which the curved tracks below lead to the cableway tower. To the west of battery C, preparations are being made for two more batteries. (MWRD photo 10963.1)

Photograph 4.55: Nearly a year later on June 18, 1925, battery C is completed and construction is well along for batteries B and A to the west. The operating galleries and final settling tanks are north of the aeration batteries. After the plant was placed in service, the naming of the batteries was reversed, with A to C from east to west. Battery D was added to the west in 1962. (MWRD photo 11839.1)

Photograph 4.56: A manual gantry crane is used to position materials, piping, baffles, and precast air diffuser plates in the bottom of the aeration tanks on April 14, 1925. The District experimented with different materials for diffuser plates and manufacturing techniques. The plates used were precast on site in a separate building before installation. (MWRD photo 11669)

Photograph 4.57: July 6, 1928. Looking down at the interior of an aeration tank, showing air diffuser plates in place. Air is piped through the wall to several locations, where connecting pipes protrude as shown and convey the air into the air diffuser plate box through the inverted "U" shaped piping. A valve with an extended vertical stem controls the flow of air. (MWRD photo 14121)

Photograph 4.58: Aeration tanks in service on April 15, 2015. The piping for air delivery to the diffusers has changed, but the process is the same. Further, present technology monitors the dissolved oxygen in the aeration tank and adjusts the flow of air to optimize the process. The aeration tank is the most crucial step in the process of reclaiming water via the activated sludge process. (Photo by author)

Photograph 4.59: Four engineers were inspecting the air mains in 1965 and died from suffocation due to a lack of oxygen caused by excess methane. As a result, the underground air mains were abandoned and replaced with twin overhead air mains in 1968. One main, shown here on April 15, 2015, continues west to batteries C and D. The other main terminated at batteries A and B. (Photo by author)

Photograph 4.60: September 2, 1924. The final settling tanks were under the north end of the cableway with walls also constructed with prefabricated steel reusable forms. They were square to make efficient use of the space available. Each lower corner was rounded for efficiency of the rotating raking mechanism and the floor sloped to the sump in the center of the tank. (MWRD photo 11141)

Photograph 4.61: Laborers are finishing the concrete floor of the final settling tank on July 21, 1925, with the raking mechanism in place for a smooth contact surface. The aerated liquid from the aeration tanks enters the top of the tank along the wall and the effluent is drawn off by the channels above the raking mechanism. The rake arms rotate slowly and the blades on the rake arm bottom move the settled solids, called sludge, to the center sump for removal from the tank. (MWRD photo 11991)

Photograph 4.62: Final tanks looking southwest on June 7, 1928. Originally, ten square final settling tanks were built for each battery. Additional circular final clarifier tanks were added to each battery in 1937 and 1988, the latter to meet more stringent suspended solids effluent limits required by the Clean Water Act. Effluent from the final tanks is piped to the effluent conduit. (MWRD photo 14035)

Photograph 4.63: A typical circular final clarifier tank on April 15, 2015. The aerated liquid from the aeration tanks enters the final tank near the bottom center, flows vertically and then horizontally toward the effluent trough on the tank perimeter. In the radial flow pattern, the velocity diminishes facilitating solids settlement. The circular tank design is more operationally efficient than square tank design. (Photo by author)

Photograph 4.64: Laborers are pouring concrete into the wood forms for the crown of the effluent conduit north of the final settling tanks on December 11, 1923. The effluent conduit ran the length along the north side of the plant site and discharged into the North Shore Channel. (MWRD photo 10376)

Photograph 4.65: Construction of the effluent outfall looking northeast on May 11, 1926. The single effluent conduit diverges into five passages at the outfall to reduce the turbulence in the channel. The effluent flows south to the North Branch, but because of temperature and chemical differences between the effluent and water in the channel, when the upstream channel is stagnant, a temporal density current can impact the channel as far north as Main Street. (MWRD photo 12668)

Photograph 4.66: The channel cofferdam remains in place on July 9, 1926, but has been breached to allow water into the outfall and effluent conduit. This photograph, taken from the electric railroad bridge, also shows in the background construction of the pump and blower building at left of center and preliminary settling tanks to the right of center. (MWRD photo 12808)

Photograph 4.67: July 30, 1928. The Service Building contains the central boiler plant, maintenance shops, and a storeroom. The tower contains a 150,000-gallon water tank providing an adequate supply and consistent pressure for the entire plant. At the time this plant was built, water was obtained through a dedicated water main from a City valve vault at Kedzie and Touhy Avenues. The chimney, an imposing landmark for several years, was reduced in height in 1992. (MWRD photo 14199)

Photograph 4.68: July 30, 1928. The electrical substation in a separate building was outfitted to use electricity from the Lockport Powerhouse via the transmission line to Wilmette and also from Public Service Company of Northern Illinois transmission lines. Power from Lockport was phased out with the reduction in Lake Michigan diversion after 1930. (MWRD photo 14204)

Photograph 4.69: Under construction on April 15, 2015—and due to be placed in service in 2016—is the disinfection facility for an added process to meet a new effluent limit for fecal coliform bacteria. The process uses ultraviolet (UV) radiation to inactivate both good and bad bacteria and other pathogens in the final effluent. In the building at right, high intensity bulbs emit UV light as final effluent passes around the bulb. (Photo by author)

Epilogue

The O'Brien plant sewershed is one of seven major sewersheds managed by the District, and it is not isolated. Through the intercepting sewer system, waste sludge pipeline and Deep Tunnel, the O'Brien plant is interconnected with other plants. Although each plant has a separate permit issued by the Illinois Environmental Protection Agency, a careful reading of the permits reveals that they all are similar in limits imposed, conditions for compliance, monitoring, and reporting; also, all include canal system or receiving waterway monitoring and reporting. In addition, a consent decree governs the discharge of combined sewer overflow. In other words, the four District water reclamation plants discharging to the canal system, the canal system itself and the Deep Tunnel system is basically regulated as one integrated environmental system.

The North Branch and North Shore Channel, only two of many canal system reaches, together total 15.4 miles in length within the total 77.1-mile canal system. The O'Brien sewershed depends on the canal system to return reclaimed water to the environment and, at numerous locations along the length of the two reaches, to receive overflows from the sewershed. Although the canal system was originally constructed to dilute and carry away sewage, its purpose for drainage remains. To the present time, it continues to be an integral part of the infrastructure for management of the water environment. These two reaches are not separate from the intercepting sewers, pumping stations, O'Brien plant, and Deep Tunnel—all are an integrated, interdependent system.

Reference is made in Appendix A-4 to the American Society of Civil Engineers' 1954 designation of the District as one of the seven wonders of modern American engineering. The society recognized all the District's infrastructure—intercepting sewers, pumping stations, plants, and canal system—as one integrated engineered wastewater system. Again, in September 2001, the society recognized the District's infrastructure, this time including the Deep Tunnel, as a Monument of the Millennium.

Constructed Canal System

Misunderstandings of the purpose and role of the canal system often result in false characterizations. I have heard the reversal of the flow of the Calumet and Chicago Rivers referred to as a "colossal blunder" and the canal system referred to as a "highway of environmental destruction" and a "deeply flawed system." Little or no explanation is given aside from these sound bites, nor are technically sound alternatives offered. In addition to protecting the quality of Lake Michigan, the canal system provides efficient drainage and avoids having pumping stations and treatment plants right on the lakefront.

Reversing the direction of river flow and building the canal system was inevitable. The Citizen's Association, a strong and influential advocacy group of its day, was a leading proponent of reversing the flow of the Chicago River to solve the critical threats of flooding, communicable disease, and the river nuisance condition. Eliminating the other discharges of sewage to the lake followed and, except for the Calumet River, was completed by 1907. In the 1909 Plan of Chicago, better known as the Burnham Plan, central goals of city planning and development were reclaiming the lakefront for the public and foreclosing any notions of using the lakefront or the lake for sewage or stormwater disposal. By 1933, the state and federal government completed the Illinois Waterway using the District's canal system as the connection to Lake Michigan. Diversion of water from the lake was made a part of the federal navigation project, just as the U.S. Supreme Court justified in the 1930 Decree.

With federal and state involvement, the District's original canal system was adapted to accommodate commercial navigation. More recently, with the Clean Water Act and federal and state environmental regulation, the District has improved the quality of water reclamation plant effluent, Deep Tunnel has lessened the impact of combined sewer overflow, and supplemental aeration has further impoved water quality, thus promoting recreation on many reaches of the canal system. However, the canal must be taken for what it is—an artificial, man-made drainage system to serve the metropolis. Various reaches have different names but they are all canals, either man-made canals where no natural channel previously existed, or natural channels that have been modified to such an extent as to eliminate most natural attributes. The movement of water is artificially controlled by numerous hydraulic structures.

The effluent discharged from four District water reclamation plants and numerous small plants throughout the entire watershed make up over three-quarters of the flow leaving the canal system at Lockport. In regulatory parlance, the canal is *effluent dominated.* The balance consists of direct diversion from Lake Michigan for lock operation and water quality maintenance, flow from tributary streams, and direct discharge of stormwater runoff. Little, if any, is what could be considered flow from a natural watershed, since the watershed is almost fully developed. During the cool and cold weather months, flow in most reaches of the canal system is nearly all effluent; during warmer months, flow is about half effluent and half direct diversion and stormwater runoff.

Some have a notion to undo the flow reversal and/or separate the watersheds, but this is simply not feasible. The Chicago area has been engineered to drain down the Illinois Waterway, and a change would require re-engineering the drainage system at considerable cost, be less sustainable, and result in unintended consequences.

Aquatic Habitat

Except for a few stagnant reaches, the canal doesn't freeze in the winter. Discharge rates are markedly steady, lacking a natural pulse caused by diurnal, seasonal, and weather changes. At elevated flow the canal does not go out if its banks, so floodplain areas are nonexistent. The slow-moving, unnaturally-deep canal lacks pools, riffles and runs characteristic of natural rivers, making the canals habitat-challenged. Tree cover is limited to a few reaches. Canal banks are composed of vertical walls of concrete or steel, or steep rocky slopes. Unnatural water depths and flows are artificially imposed by hydraulic structures such as navigation locks, a powerhouse, pumping stations, and water reclamation plants. Two-thirds of the system is used by commercial navigation that produces forceful wave washes that prevent or severely restrict development of shoreline habitat.

The ambitious goal of the Clean Water Act, "... to restore and maintain the chemical, physical and biological, integrity of the Nation's waters...," is a tough challenge for the man-made canal system. Although federal regulations allow for such modified or unnatural waterways to be classified differently than natural rivers, humans are prone to attempt to make the canal system into something it isn't. Habitat in the canals is characteristically void or limited. Most aquatic organisms that exist there are species that are generally tolerant of conditions in the canal system. Technology can be employed to achieve compliance with chemical water quality standards, but for a waterway that didn't exist in nature or has been irreversibly modified, restoring biological and physical integrity is not practical. And creating habitat is challenging—there are no success stories of making canals act like rivers. Think of the canal as an aquarium: the water is from the tap, the foliage and bottom material are artificial, a bubbler is needed to maintain dissolved oxygen and induce circulation, and temperature regulation is necessary.

Perhaps the best of what one might call habitat in the canal system is found in the North Shore Channel, upstream of the outfall of the O'Brien water reclamation plant. The outfall is located 1,000 feet north of Howard Street. During November through April, water in the four-mile reach up to Wilmette is frequently stagnant. Unlike other parts of the canal system, thick ice cover forms during winter. During the balance of the year the water moves slowly if lake water is being diverted at Wilmette. Tree cover is lush and power boats are rare, reducing the severity of channel bank wave wash. Normal depths are not excessive and stormflows cause more frequent variability in depth. The soft clay sediments and prevalence of bottom feeding common carp often result in high turbidity. Despite this last drawback, District aquatic biologists report finding some evidence of fish reproduction for species adapted to nesting in silty bottom areas.

Sediments in the canal are contaminated, but mostly not so seriously that remediation is required. For canal segments with sustained flow, sediments are disturbed by occasional prop wash and are moved along by periodic storm flow. The more seriously contaminated areas are in the off-channel areas such as the North Branch immediately south of Diversey Boulevard, the North Branch turning basin immediately south of North Avenue, and the North Branch Canal. In these locations, where the water velocity is lower, deep sediments have accumulated over many decades. The high oxygen demand of the sediments depletes the dissolved oxygen in the overlying water, impairing compliance with the dissolved oxygen water quality standard. A technology called *active capping* has been successfully demonstrated in marine waters to remediate contaminated sediments. This technology can be applied in the District's canals, but until that happens, water quality in the canal will continue to be impaired by wide area and off-channel sediments. Alternatively, these areas could be separated with a physical barrier and the area partially filled to create a wetland. The wetland would provide water quality benefits while not diminishing the flood storage capacity of the canal. The barrier would serve to stabilize the sediment and deflect floating and wind-blown trash.

Although fish are collected in the canals, species richness and diversity is low, and the canals generally lack suitable habitat for fish reproduction. The canals can be made more fishable by seasonally stocking sport fish from a hatchery. The 70 species of fish in the canals demonstrate that a variety of fish species can survive despite periodic episodes of low dissolved oxygen. However, limited habitat favor the more tolerant fish species. The common carp, bottom feeders, stir up sediment, which makes survival difficult for sport fish that feed by sight.

Canal System Ecological Improvement

In 2013, the District began participation in the Chi-Cal Rivers Fund, a private-public partnership funding the creation and restoration of aquatic habitat, increased public access to the waterways and green infrastructure projects. The project is administered by a third partner, the National Fish and Wildlife Foundation, and projects can be implemented anywhere in the Calumet and Chicago River watersheds. This is an excellent example of the District taking responsibility for water quality and ecological improvement of the waterways in its jurisdiction. The District is the only governmental agency that has the historic precedent, hometown interest, and specific statutory authority for this responsibility.

With the cooperation of the Illinois Department of Natural Resources, the Friends of the Chicago River (Friends) released tens of thousands of hatchery catfish in the North Shore Channel, North Branch, and other parts of the canal system. This was the initiation of a project to improve habitat by deploying nesting cavities in the canal bottom for the catfish to breed successive generations. Monitoring the results over time will show if this is a successful technology for man-made canals. Implementation of creative habitat improvement ideas may lead to successful technologies that improve ecological habitat in the canal system.

Dissolved Oxygen

Aquatic life requires oxygen, and fish, for example, are able to absorb dissolved oxygen from the water; hence there are water quality standards for levels of dissolved oxygen to ensure survival of fish and other aquatic organisms. Natural streams replace the dissolved oxygen consumed by aquatic organisms and decaying vegetation by reaerating themselves, but the canal needs help. Treated effluent discharged from the O'Brien water reclamation plant is rich in dissolved oxygen, but as water flows slowly downstream, the dissolved oxygen is consumed by aquatic life, suspended organic matter, and organic sediments. In order to maintain compliance with the state's guidelines for dissolved oxygen, the channel requires the dissolved oxygen to be supplemented artificially. The District's Devon Avenue and Webster Avenue Instream Aeration stations, built in 1978, both use diffuser plates on the channel bottom along the side walls to infuse the channel with tiny bubbles.

At Devon Avenue, air compressors are located in the building to the west of the channel and south of Devon. From the bridge's south railing, one can see the building and the rising curtain of air bubbles along both the east and west sheet pile channel walls. At Webster

Avenue, the air compressors are neatly tucked into an obscure building along Ashland Avenue south of the Webster Avenue Bridge. From the bridge's south railing, one can see the building and the rising curtain of air bubbles along the west bank. The diffusers on the east bank are located north of the bridge. The compressors rarely need to run in the winter weather because cold water holds more dissolved oxygen, but the compressors run nearly continuously spring, summer, and fall. These two installations may not look elegant, but they are efficient at replenishing dissolved oxygen.

The responsibility of maintaining compliance with the dissolved oxygen water quality standard rests largely with the District. Consider this language in the discharge permit issued by the Illinois Environmental Protection Agency for the O'Brien plant outfall and 46 combined sewer overflow outfalls in the O'Brien plant sewershed controlled by Deep Tunnel: "…discharges from the outfalls…shall not cause or contribute to violations of applicable water quality standards or cause or contribute to designated use impairment in the receiving waters." Any dissolved oxygen standard non-compliance in the North Branch or North Shore Channel points to the District as causing or contributing to the non-compliance. In addition to the two supplemental aeration stations mentioned above, more may be necessary in the future to assure compliance with more stringent dissolved oxygen water quality standards adopted by the Illinois Pollution Control Board.

Direct discretionary diversion from Lake Michigan is also used to maintain compliance with water quality standards, however, the amount is limited and may be distributed for other reaches of the canal system.

Deep Tunnel Future

Until the McCook Reservoir is in service to significantly reduce the frequency and volume of combined sewer overflows, occasional overflows will continue to cause periodic depressions in dissolved oxygen concentrations. Even after this reservoir is in service, less frequent overflows may occur and supplemental aeration will continue to be needed simply because the canals cannot reaerate themselves without help. The full 12-billion-gallon storage capacity of the Mainstream and Des Plaines reservoir/tunnel system is only available when the system is empty. If it is full following a large storm or a prolonged rainy period, the system will be of little or no benefit until it is partially or completely dewatered. The system is dewatered by pumping at the Mainstream Pumping Station in Hodgkins through a five-mile tunnel to the Stickney water reclamation plant.

The Stickney plant serves a large area; following a large storm or rainy period it can take two or three days for the plant to return to normal dry weather conditions. As the plant is in this return mode, dewatering of the reservoir/tunnel system can commence within maximum plant capacity. The maximum pumping capacity of the Mainstream Pumping Station is such that it would take 22 days to empty the system. Thus, adding a two- or three-day return period, it can take about 25 days for the plant to return to normal and empty the reservoir and tunnels. If there is more rain in that 25 day period, the length of time to dewater the reservoir and tunnels will be extended. However, in operating the reservoir/tunnel system, the District may close gates after a storm to isolate the reservoir and begin emptying the tunnels. When the tunnels are near empty, pumping from the reservoir will begin. While the reservoir is full or nearly full, the degree of combined sewer overflow reduction and flood relief will be about the same as it is at present, with only tunnel storage available. When the reservoir is isolated, any available capacity can be utilized by manipulating gates and valves, thus maximizing all available capacity in the reservoir/tunnel system.

Canal Waterfront and Recreation

Some believe the canal can be a second waterfront comparable to Lake Michigan. The canal can qualify as a waterfront, but it is unreasonable to compare it to the lake. On the canal there are few open areas, no beaches, and no mile-upon-mile of access. Extensive private ownership limits access to the canal bank and water edge. By contrast, most of the 1.5-mile reach of the Chicago River is open to the public. Both the entire south bank and north bank east of Michigan Avenue enjoy a continuous canal walk atop the dock wall. The north bank canal walk west of Michigan Avenue is intermittent, sometimes atop the dock wall or at street level.

Some steps are certainly possible to make parts of the canal attractive: the banks can be beautified, paths can be installed for biking or walking, water access can be made available for boating, and various cultural amenities can be included for public enjoyment. Boating can be enjoyed throughout the canal system, but caution is advised for deep depths, unpredictable currents, and lack of safe places to exit the water should a boat capsize.

Plentiful opportunities exist for recreation along the North Branch and North Shore Channel. There are several parks between Lake Street and Lawrence Avenue in the city along the 7.7-mile-long North Branch, and more, on land owned by the District, along the 7.7-mile length of the North Shore Channel from Lawrence Avenue in the city to Sheridan Road in Wilmette. (See Tables 4 and 5.) A continuous trail runs from Lawrence Avenue to Green Bay Road.

Swimming

Some envision a future when swimming in the canal is possible. However, unless it is on a limited scale and liability issues are properly addressed, it is unlikely that any public agency will step forward to take responsibility. State regulations for public bathing beaches are stringent, and the financial liability, should anyone be injured or lose their life, could be staggering. At present, swimming is not advised due to limited access, deep depths, lack of lighting at night, vertical walls, unpredictable currents, and no lifeguards. In 2003, the Illinois Environmental Protection Agency directed the District and other municipalities to install signs on their respective canal properties to warn people of the threat to health from contact with non-disinfected water. But even with forthcoming disinfection of water reclamation plant effluent, the threat will continue because there are many sources of bacteria in the canal system and tributary areas. If swimming is desired, Lake Michigan beaches or swimming pools are much safer places.

The forthcoming disinfection at the O'Brien plant is not the first time that reclaimed water discharged to the canal system will be disinfected. At the four District plants discharging to the canal system, disinfection was practiced from 1972 through 1983 under a state rule; that rule was eliminated in 1984. During this period disinfection was effected using sodium hypochlorite, which is similar to household bleach. However, the old rule didn't require dechlorinating; hence, to meet the effluent limit there was a surplus of the chlorine ion in the discharge to the canal and that surplus was toxic to aquatic life. In other words, while disinfection was killing some microorganisms and pathogens in the plant effluent, it was also killing the fish in the canal.

After the practice was ceased in 1984, fish came back. However, prior to 1984, the chlorine didn't kill the midge fly larvae in the North Shore Channel sediment, and the flies were a serious nuisance along the canal. The District attempted to reduce the midge fly infestation through aerial spraying for adult midges and applying larvicide in the canal. Once chlorination was discontinued at the O'Brien plant and fish life was sufficiently restored in the canal, the midge fly nuisance disappeared; the fish were feeding on the midge larvae.

Many organizations, the District included, talk about the tremendous improvements in canal water quality brought about by improved treatment technology, plant expansion and the Deep Tunnel. The increase in the number of fish species in the canal is held up as an example. Indeed, water quality has shown improvement, but the fish increase was due to removing the toxicity of chlorine more than water quality improvement. For over two decades the number of fish species has plateaued at about 70, but common carp and carp hybrids are still the most abundant, and this is not likely to change. Common carp and urban canals are well suited for each other.

Disinfection will again be practiced as a matter of policy at the O'Brien water reclamation plant. Numerous environmental advocacy organizations, elected officials, news media, and even the U.S. and Illinois Environmental Protection Agencies began a campaign in 2003 voicing strong support for disinfection. The District took the matter under consideration, conducting research on the actual health benefits of disinfection in the canal environment where the microbiology is significantly different than at fresh and marine water beaches.

With improved canal water quality came increased use of the canal for recreation, and no records of reported illness were available to justify the implementation of disinfection. The District conducted a three-year epidemiological study involving 11,000 participants who engaged in boating recreation in locations including the canal system, Forest Preserve District lakes and rivers, and oither nearby lakes and rivers. The results showed no significant difference in the incidence of illness between those engaging in boating recreation on the canal system versus other waters. Thus, there would be no increased public health benefit by implementing disinfection. However, in 2011 the District commissioners decided, as a matter of policy, to implement disinfection. The decision was eased by a May 2011 letter from the U.S. Environmental Protection Agency to the Illinois Environmental Protection Agency stating, in essence, that it deemed the canal not only suitable for "incidental" contact such as boating and recreating, but for "primary" contact as well, which includes swimming.

The federal bacterial criterion for swimming is based on risk; eight or fewer illnesses per thousand swimmers is considered safe. But even if effluent is disinfected, a swimmer immersing in it is taking a risk. Since canal recreationists are not required to report certifiable illness, there will be no evidence to demonstrate that disinfection is beneficial. However, since disinfection gives comfort to those who think it helps, perhaps this is sufficient benefit. For the many who recreate on the canal without concern for the fecal coliform count, their pleasure is not likely to change.

Stormwater Management

Managing stormwater involves two capacities, conveyance and storage, the latter to temporarily hold water until it can be safely released via the former. Traditionally, structural conveyance and storage facilities have been the norm. To the extent stormwater can be evaporated or infiltrated, that can lessen the capacity needed for conveyance. However, with intense rainfall little time is available for significant reduction in stormwater runoff. The Deep Tunnel is intensely structural, with drop shafts and tunnels for conveyance; the tunnel itself has some storage and there are large reservoirs for additional storage. Concepts like best management practices, low impact development, or green infrastructure currently stress holding stormwater runoff near where the rain falls to reduce the size of large structural conveyance and storage facilities. Regardless of technology, the concepts of conveyance and storage apply.

Two recent examples in the north area are illustrative. The Village of Winnetka considered a tunnel to convey excess stormwater from the western part of the village to Lake Michigan. Little storage was necessary and nearly 50 feet of elevation difference made this attractive as no pumping was necessary. However, large estimated costs have led the Village to defer the tunnel concept while considering storage alternatives using the Skokie River as the outlet. The Skokie River is the current outlet for Village storm sewers, but pumping is required due to the flat landscape. The search for sufficient alternative storage to provide the same level of flood relief as the tunnel will be challenging.

The Albany Park neighborhood, along the North Branch upstream of the North Branch Dam, has experienced severe flooding due to the restricted capacity of the North Branch. Although the neighborhood is served by combined sewers and the Deep Tunnel provides relief for local drainage, the floodwater invading the community from upstream results in overbank flooding. Providing storage would be expensive; there are limited opportunities upstream, and neighborhood development occupies former floodplain areas. Purchasing private flood-prone property to provide some storage and increasing channel capacity were not considered acceptable options. Instead, the choice was a tunnel so floodwater would bypass the neighborhood. The tunnel takes in excess floodwater west of Pulaski Avenue, runs east under Foster Avenue, and discharges to the North Shore Channel.

When the Chicago area is awash in stormwater runoff, many believe that opening the flood gates to Lake Michigan will eliminate or reduce flooding in their local area. This is true only to a limited extent, and only for areas in close proximity to the canal system. Floodwater can reach the canal system only via tributary streams and combined or separate storm sewers. For communities that are distant from the canal system, stormwater would have to travel long distances over long time periods; it is usually the limitations in local drainage capacity that cause local flooding.

As a last resort, floodwater is released to Lake Michigan by the District at some or all of the three lakefront control structures on the canal system. This release happens when there is too much floodwater in the system; even if the floodwater were not released intentionally, it would flow over the top of the walls separating the canal from the lake, resulting in consequences more serious than from a controlled release. Releases from the North Shore Channel at Wilmette are more frequent, because this location is the most distant from the canal system outlet at Lockport and the channel has less capacity than other parts of the canal system. In the 26-year period from 1990 through 2015, there have been 23 events when floodwater was released to the lake. In 22 of these events, release was made from the North Shore Channel; in 13 events, release was made from the Chicago River; in five events, release was made from the Little Calumet River to the Calumet River.

When the gates are opened, they remain so for a short time—just enough time to return the canal system to water levels that won't result in overbank or local flooding along the canals. For the 22 releases mentioned above for the North Shore Channel, only six were longer than 12 hours. In mid-September 2008, two storm fronts converged on the Chicago area and covered Cook County with five to eight inches of rain over a three-day period. The amount of rain and the large area covered was unprecedented and resulted in discharge to Lake Michigan at all three lakefront control points for three days, a record-breaking flood. Based on a 25-raingage network, this storm exceeded eight inches at three locations and seven inches at ten locations. The areal extent is what made this storm remarkable.

References

google.com/maps.

Illinois Department of Public Health. Environmental Health Protection. Swimming Facilities. Dph.illinois.gov/topics-services/environmental-health-protection/swimming-facilities.

Illinois Pollution Control Board. Rulemaking for Use Classification and Water Quality Standards. Docket R2008-009, Sub-dockets A through E.

Metropolitan Water Reclamation District of Greater Chicago. Proceedings of the Board of Commissioners, 2011 through 2015.

mwrd.org.

Rijal, Geeta. Personal communication. January 2016.

Stoner, Nancy, Acting Assistant Administrator for Water for the U.S. Environmental Protection Agency. Letter dated May 11, 2011, to Lisa Bonnett, Interim Director of the Illinois Environmental Protection Agency.

"Thornton Composite Reservoir Greatly Boosts Chicago's CSO Storage Capacity." *Civil Engineering Magazine*. (October 2015): 34.

Wasik, Jennifer. Personal communication. December 2015.

Westcott, Nancy E. Continued Operation of a 25-Raingage Network for Collection, Reduction, and Analysis of Precipitation Data for Lake Michigan Diversion Accounting: Water Year 2008. Contract Report 2009-04. Champaign, IL: Illinois State Water Survey, February 2009.

wikipedia.org.

Table 4: Recreation Sites along the North Branch

Park/Site	Bank	Location	Owner	Water Access
California Park	West	South of Irving Park Road	Chicago Park District	No (1)
The Garden Dirt Jumps	East	North of Belmont Avenue	Chicago Park District	No
Goose Island Overlook	West	Elston Avenue South of Division	City of Chicago	No
Henry Horner Park	West	South of Montrose Avenue	Chicago Park District	Yes (2)
Jimmy Thomas Nature Trail	East	North and South of Diversey Parkway	Lathrop Homes	No
Leland Park	East	Leland Avenue	Chicago Park District	No
North Branch Turning Basin Overlook	East	North end of Goose Island	Wrigley Innovation Center	No
Richard Clark Park	East	South of Addison Street	Chicago Park District	Yes
Riverbank Neighbors Park	East	South of Montrose Avenue	Riverbank Neighbors (3)	Yes
Sunnyside Park	East	Sunnyside Avenue	Chicago Park District	No
Ward Montgomery	East	Erie Street	Chicago Park District	No
Webster Wildlife Site	West	South of Webster Avenue	African Wildlife Foundation	No

Notes:

(1) Swimming pool in park

(2) Under construction in 2015

(3) Under permit to NeighborSpace

Table 5: Recreation Sites along the North Shore Channel

Park/Site	Bank	Location	Lessee (1)	Water Access
Ronan Park	Both	North of Lawrence Ave. to Argyle Street	Chicago Park District	No
River Park	Both	Argyle Street to Foster Avenue	Chicago Park District	Yes (2)
Legion Park	Both	Foster Avenue to Peterson Avenue	Chicago Park District	No
Park No. 526	Both	Peterson Avenue to Devon Avenue	Chicago Park District	No
Park No. 538	East	Devon Avenue to Touhy Avenue	Chicago Park District	No
Lincolnwood Centennial Park.	West	Devon Avenue to Touhy Avenue	Village of Lincolnwood	No
North Shore Sculpture Park and Trail	West	Touhy Avenue to Emerson Street	Skokie Park District	No
Dammrich Rowing Center	East	North of Oakton Street	Skokie Park District	Yes
Trail and Soccer Field (Channelside Park)	East	South of Main Street	Skokie Park District	No
Harbert Park	East	Main Street to Dempster Street	Evanston Recreation Department	No
Beck Park	East	Church Street to Emerson Street	Evanston Recreation Department	No
Butler Park	East	Emerson Street to Bridge Street	Evanston Recreation Department	No
Ladd Arboretum	West	Emerson Street to Green Bay Road	Evanston Recreation Department	Yes (3)
William H. Twiggs Park	East	Bridge Street to Green Bay Road	Evanston Recreation Department	No
Canal Shores Golf Course	Both	Metra Union Pacific / North Line to Sheridan Road, Wilmette	Evanston Wilmette Golf Course Association (4)	No

Notes:

(1) All sites are on property owned by the district

(2) Water access is available on the west side of the channel near the North Branch Dam and a swimming pool is on the east side of the channel

(3) Water access adjacent to the Evanston Ecological Center

(4) Subleased from the city of Evanston and Wilmette Park District

APPENDICES

Appendix A-1: Legislation

ACT OF 1889

The Act of 1889 not only enabled the creation of the Sanitary District of Chicago by referendum, but also provided elected officials with the authorities to construct the channel and make other improvements to reverse the flow of the Chicago River to protect the public health of citizens in the District who were receiving water from the lake. As explained in *Building the Canal to Save Chicago*, pages 269, 335, and 338, successive legislation was necessary in 1893, 1895, and 1897 to authorize additional tax resources and police power and to clarify navigation requirements. Beside these, other amendments prior to 1900 strengthened the requirement for dilution and authorized the Illinois attorney general to take enforcement action against the District upon a complaint from another unit of government that the District was violating the statute. Some of these provisions were undoubtedly added to gain downstate support for passage of the District-sponsored legislative initiatives.

In the years since, numerous amendments to this original act and other legislative acts affecting municipalities have governed and guided the activities of the District under its original and successor names. The Acts of 1901 and 1903 are pertinent to the work of the District explained herein.

ACT OF 1901

Retiring President Boldenweck drew attention in his annual message on December 3, 1900, to the continuing problems of sanitation in the areas beyond the District borders yet within Cook County. He suggested that the areas be annexed into the District or a means be found for the drainage from these areas to be conveyed to the Sanitary & Ship Canal and away from Lake Michigan. The following day at the annual board meeting a committee-of-the-whole was formed to address the issue and seek annexation through an act of the General Assembly. Legislation was drafted and introduced early in 1901.

In April 1901, the District was served with a petition signed by a committee of twelve appointed by the City of Evanston objecting to the proposed legislation for annexation of areas to the District. The petitioners were of the opinion that the lakefront areas of Lake County should be included with the northern suburbs in Cook County to give the District total control of Lake Michigan in Illinois. They were opposed to the additional tax imposed on annexed areas to pay to retire the bonds that funded past construction expenditures. In their opinion, the additional tax was unconstitutional and would lead to endless litigation and poor relations. The petitioners requested that the District communicate with the state senate leadership and request that a public hearing be held in the affected areas before bringing the bill to a vote. The District initiative for annexation of adjacent areas didn't secure sufficient support and was not enacted.

Two other bills before the general assembly dealing with local government finances could have adversely impacted the District. By working with the legislators and other interests, one objectionable bill was never enacted and the other more favorable bill was amended to clarify language regarding condemnation of property for river improvement, replacement of obstructive bridges with modern bridges, and improvements for the river channel to pass the required flow volume; it also allowed municipalities to own, operate, and maintain bridges built by the District. This act gave the District clear authority to replace many obstructive bridges over the Chicago River and South Branch and use the channel of the North Branch for drainage improvements authorized under the Act of 1889. The Act of 1901 also clarified the responsibility for operation and maintenance of bridges built or replaced by the District, allowing municipalities to assume this responsibility.

ACT OF 1903

The proposed annexation legislation was again introduced in 1903, after outside counsel was sought and numerous representatives of the two adjoining areas to be annexed were consulted. The District also sought to clarify the authority to develop water power and increase taxing authority to pay for it. By separating the two proposals into two draft bills, the trustees hoped that success could be achieved with less complexity. However, both bills came before one legislative committee and numerous objections were raised and amendments introduced. A delegation of citizens from the North Shore suburbs presented petitions on the same concerns as outlined above. The Chicago Board of Local Improvements supported the annexation of the adjoining areas to end the pollution of Lake Michigan. The Illinois & Michigan Canal Commissioners addressed their concerns regarding water power by proposing several amendments. To clarify any misunderstandings about what annexation would accomplish, the District submitted a detailed statement to the governor and all members of the general assembly. The committee combined both bills into one, and the revised bill passed the General Assembly and was approved by the governor in May.

Effective July 1, 1903, the act was titled *An act in relation to the Sanitary District of Chicago, to enlarge the corporate limits of said district and to provide for the navigation of the channels created by such district and to construct dams, water-wheels and other works necessary to develop and render available the power arising from the water passing through its channels and to levy taxes therefor*. This act was the most far-reaching statutory authority since the Act of 1889 enabling the creation of the District. Section 1 extended the corporate limits of the District by annexing territory to the north and south, doubling the area of the District. (See Figure A-1.)

Section 2 authorized the District to provide for the drainage of the annexed territory by constructing one or more channels, use the Illinois & Michigan Canal Calumet feeder and lands adjacent to the feeder belonging to the state of Illinois for constructed channels, and construct a channel across the Illinois & Michigan Canal without having to restore the canal or its feeder. Section 2 required, among other details, the District to:

- install gates where the constructed Calumet channel connects to the Calumet River,
- construct a navigation lock and a navigable channel from the current channel at Lockport to the Upper Basin in Joliet before construction of the Calumet channel,
- require protective redundant gates if only one navigation lock is built, and
- require construction of a suitable dock and navigation terminal for use by the Illinois & Michigan Canal Commissioners.

The authority in Section 1 and part of Section 2 was sought by the District. Many of the requirements in Section 2 were added at the request of the Illinois & Michigan Canal Commissioners to protect their interests.

Section 3 required the District to allow all water craft to navigate its channels without charge and to obey federal navigation rules. Section 4 prohibited the District from imposing any special assessment or tax upon property in the annexed area to defray the cost of channels and other works constructed prior or subsequent to annexation. This section was added at the request of the areas to be annexed and as a deterrent to the referendum option in Section 9. Section 5 authorized the District to develop water power north of the Upper Basin in Joliet, effectively prohibiting the District from developing water power downstream of the Lockport area. Section 6 allowed the electricity to be transmitted to various cities and towns within the District for municipal purposes or to others.

Section 7 authorized the District and the county clerk to impose an additional tax limited to three years and amounting to 0.25% of the value of the taxable property. This section provided the needed additional financial resources for channel construction in the annexed area and the development of water power. Section 8 required the District to comply with federal law pertaining to abandoned land of the Illinois & Michigan Canal. Section 9 allowed for opposition to annexation by the voters in the area to be annexed. Within 60 days, if more than three percent of the voters in the annexed area requested a referendum, the referendum would be in the general election in November 1904. Annexation would not occur until a majority of the voters approved.

Appendix A-2: North Branch Canal

The city's famed and infamous Goose Island was another human-made creation, not the result of natural forces. Details are sketchy and hard to pin down, but in the 1850s, business interests related to the City's first mayor, William B. Ogden, were mining clay for the manufacture of bricks. It is said that the resulting excavation became the Ogden Canal, but this odd-shaped clay pit was different than most other clay pits in the Chicago area. However, the long and narrow clay pit/canal allowed Ogden and his pals to enhance the value of land by providing water frontage in this busy port city. While the river channel courses through a gradual bend between Chicago and North Avenues, the clay pit/canal is a mile-long straight channel connecting to the North Branch at each end, creating Goose Island. (See Figure 6).

Apparently some thought the Ogden Canal too narrow and saw an opportunity to have the District come to their aid. While busy improving the North Branch farther north, a petition was received in February 1905 from citizens, vessel boat captains, boat owners, and navigation agents from a number of Great Lakes cities in addition to the city, requesting widening of the Ogden Canal from Halsted Street to its junction with the North Branch. The expressed concern was the adequacy of capacity and navigability when additional flow would be introduced via the Lawrence Avenue conduit, then under construction. In his report on the matter the chief engineer stated that with all sources considered, including the Fullerton Avenue flushing tunnel, Lawrence Avenue conduit, and the North Shore Channel, the rate of flow would not exceed 1,830 cfs and have a velocity of 0.42 mile per hour, well within what was safe for navigation and suitable for sanitary conditions. The petitioners were advised there would be no widening.

Ogden had so many features named for him that the Ogden Canal handle gradually faded from view. The canal itself began to fade from view as it became shallower with sediment deposition and much less boat traffic. The canal probably would have been filled in with one or more public works projects such as bridge building or street connections, but late in the twentieth century it was saved from extinction by recreation as canoeists and small boat pilots took more interest in it as a shorter and more tranquil passage. Converting the old railroad swing bridge at the north end of Goose Island to a walking path, building a paddle boat access at Weed Street and boat houses adjacent to Weed Street, led to a brighter future and continued usefulness for the North Branch Canal.

Appendix A-3: Fullerton Avenue Flushing Tunnel

In 1870, City Engineer Ellis Chesbrough recommended the construction of a canal or tunnel in Fullerton Avenue that would allow lake water to be pumped from the lake to the North Branch to flush the river with clean water. Conversely, the pumps could be reversed for dirty water from the North Branch to be pumped to the lake. Fullerton Avenue was the north city boundary at the time and Chesbrough suggested that in dry weather, the tunnel would allow the lower end of the North Branch to be flushed and freshened with lake water. In wet weather, the tunnel would allow excess water to be discharged to the lake, bypassing the Chicago River and downtown area. Approval was delayed due to the Chicago Fire and construction didn't begin until 1874; because of various delays the project wasn't completed until 1880.

The Fullerton Avenue Pumping Station and Flushing Tunnel consisted of a brick tunnel 12 feet in diameter and over two miles in length, two pumps located near the North Branch, and two intake/outlet structures located at the lakefront and at the North Branch. The pumps had a capacity of 250 cfs and were steam-driven with coal-fired boilers. The North Branch was improved, but as the city expanded northward, the North Branch became polluted farther north than Fullerton Avenue and the flushing tunnel became less effective. In 1889, city annexations pushed the north city boundary to Devon Avenue.

For two decades the flushing tunnel was primarily used to pump North Branch water into the lake, reducing the amount of polluted water reaching the Chicago River. The flow into the lake seemed to force lake water into the Chicago River, thereby making the water in the downtown area fresher. After 1900 and the opening of the Sanitary & Ship Canal, the tunnel was used to pump lake water into the North Branch, eliminating the polluted discharge to the lake and diluting the sewage in the North Branch. The Lawrence Avenue conduit was placed in operation in 1906, reducing the need for the Fullerton Avenue tunnel. The pumps were removed from service around 1910 and the lake intake was sealed, but the tunnel remained in service as an outlet for several sewers that had been connected to the tunnel.

Additional flushing tunnels had once been considered at Lawrence Avenue on the North Side and Thirty-Ninth Street on the South Side. However, as time progressed, needs changed and the conduits and pumping stations on these two routes were never used strictly as flushing tunnels similar to the one at Fullerton Avenue.

Appendix A-4: Name Game

The Sanitary District of Chicago was renamed the Metropolitan Sanitary District of Greater Chicago (MSDGC) by the board of trustees in 1955, attempting to impress by inflating the sound of its jurisdictional area. However, most kept referring to the MSDGC as the *Sanitary District.* The name change occurred by ordinance at a regular board meeting in October 1955 and was in effect upon passage. The ordinance stressed that nothing about the District was changed, except the name, and nothing in the transmittal letter explained the significance of the change. The redundancy of *greater* and *metropolitan* was not made clear. Since its founding, the District has been greater than the city and multi-municipal by statute.

The name change occurred a few weeks before the sixty-sixth anniversary of the establishment of the District by referendum, so the change wasn't done to celebrate age. And the change wasn't because the District was designated in 1954 as one of the seven wonders of modern American engineering by the American Society of Civil Engineers. The change was occasioned by dramatic growth. In 1955, the Illinois General Assembly increased the jurisdictional area of the District, nearly doubling its size, by adding the equivalent of nine or ten townships in the rapidly growing northwest, south, and southwest suburban areas of Cook County.

A second change was more personal; the trustees decided to relabel themselves *commissioners* because they believed the public might confuse a *trustee* with a *trusty*. This change happened in 1975 at about the same time as a scandal was brewing over a barge contract. During discussion at the board's legislative committee meeting in January 1975, the trustee advocating for the change stated "The main reason is to avoid being called a 'trusty.'" The president clarified for those attending and unaware of the different spelling: "Spelled with a 'y.'"

However, the change couldn't be made without statutory authorization and it was included in the District's legislative program for 1975; the General Assembly obliged in Public Act 79-310 approved by the governor in August 1975. Debate in the Illinois House of Representatives in June illustrates the importance of this change. Although the bill cleared committee consideration unanimously and was on the consent agenda, debate was requested. One representative took time to object to a minor name change issue taking up the valuable time of legislators. Another suggested that the change should not be made so that the District board of trustees wouldn't be confused with the Cook County board of commissioners. The bill sponsor defended the proposal saying that the trustees strongly favored being called commissioners.

The change was made without celebration or outward notice, even in the newspaper gossip columns; at the board meeting in September 1975, all were addressed as trustees and at the first meeting in October all were addressed as commissioners.

The last name change occurred in 1988; as part of the coming centennial celebration the commissioners decided to hold a naming contest. Aside from all the joke names submitted, the contest winner was the very creative Metropolitan Water Reclamation District of Greater Chicago (MWRDGC), lengthening the brand by one word. Gone was the unpleasant *sanitary* label, which always confused the public with the City department that collects garbage. In April, a simple motion passed unanimously to invoke the change, but after the meeting the attorney advised that the procedure was incorrect. At the following meeting, a motion to reconsider passed unanimously and the name change was rescinded. Later in December the name was changed by ordinance similar to the action in 1955 to be effective at the start 1989, the centennial year. Despite the change, drainage remains the District's primary responsibility and many people still refer to the MWRDGC as the *Sanitary District*.

Actually, *water reclamation* wasn't that original in 1988 as the term came up at the legislative committee meeting in January 1975; following the discussion of changing *trustee* to *commissioner*, it was suggested that the District name also be changed, and several expressed interest in *water reclamation* instead of *sanitary*. But prudence prevailed; there would be no change as concern arose that such a change might draw unwanted attention to the affairs of the District.

Of late, a number of other similar agencies have adopted the *water reclamation* modification, dropping *sanitary*. *Water resource recovery facility* is another name gaining popularity. Perhaps another name change is in store for the District as resource recovery is the latest subject of several innovative initiatives, recovering not only water, but energy and nutrients as well.

Appendix A-5: On and Attached to the Banks of the North Branch

RIVERVIEW AMUSEMENT PARK

Over 2.5 million visitors found fun and thrills annually at this memorable place along the North Branch at Belmont and Western Avenues. Riverview was an apt name because the North Branch was visible from the top of the several roller coaster rides or parachute drop, as well as from the tree-shaded picnic grove. The park opened in July 1904, while improvement of the North Branch was beginning, and was last open on Labor Day 1967. It didn't take long to disappear. Demolition began in January and was completed by March 1968. The fun and thrills have been replaced by a police department area headquarters, school, strip mall, and industrial park. The author spent many happy days at Riverview and he believes, but cannot personally confirm, that many of his classmates at Lane Technical High School played hooky there.

CHICAGO & WEST RIDGE RAILROAD

A railroad in West Ridge! Yes, you can still see the tracks from this long-gone railroad if you travel along Channel Road south from Howard Street in Skokie, just west of Kedzie Avenue, and the border with the city's West Ridge neighborhood. It served local landowners south of Oakton Street and east of what would become the North Shore Channel. At the north end, it connected to the Chicago & North Western Railroad's Mayfair Division Line just south of the Weber rail yard on Oakton Street. The south end was at Lincoln Avenue in Chicago. When the railroad was built is uncertain, but it apparently ran south along the east side of what was the Big Ditch, a north-south drainage ditch emptying into the North Branch near Foster Avenue. Libby Hill describes the Big Ditch in her book *The Chicago River: A Natural and Unnatural History*.

The railroad was probably used to transport coal to and bricks from brick manufacturers near Lincoln Avenue and Howard and Oakton Streets. A large greenhouse near Pratt Boulevard used coal for winter heating. It played a role in construction of the new channel. In 1907, the District obtained an option to purchase part of the right-of-way, thinking that the route of the North Shore Channel would cross or impinge on the railroad. However, the option was never exercised. The channel route came close, but didn't touch the railroad right-of-way; however, the District found the railroad useful in construction of the channel. Coal and structural steel were delivered over its tracks. The coal was used to fire the boilers of the construction machinery and the structural steel was used to build bridges at Devon and Touhy Avenues, and Howard Street. In 1909, the District awarded contracts for the excavation of the North Shore Channel and construction of bridges, and the railroad was probably used by the contractors for delivery of equipment and materials. The District also used the railroad directly when it was constructing Sections 10 and 11 and for miscellaneous work along the channel.

The tracks were owned by Bernard Weber, who also owned an interest in one of the brick manufacturing companies, but Weber did not own any rolling stock. The Chicago & North Western Railroad locomotives moved rail cars over the track for which Weber received a track service fee.

HOUSEBOATS

Not long after the new channel was there, houseboats appeared on the North Branch near Irving Park. Tethered to the channel bank or to one another, it was a sort of no-man's-land on water. They were there rent-free, unless whatever or whomever a boat was tethered to demanded some payment. Many occupants were registered to vote using assumed addresses on Rockwell Street. Waste disposal, solid or liquid, was easy—right into the channel designed for sewage disposal. The District ignored them, but after the North Side Sewage Treatment Works went into service in 1928, the attitude changed. The District went to court in July 1930 to oust the water squatters, charging them with pollution and obstructing navigation. The litigation failed as the District lacked proper standing. However, two houseboats weren't ignored but actually gained notoriety.

An observant citizen notified the Coast Guard in late March 1963 that a houseboat was drifting in the North Branch way up by Addison Street. The guardsmen responded in their launch and tied the unoccupied houseboat to the nearest pier. The same event occurred over the next few days. Finally, the 36-by-12-foot, rickety floating shack was towed from Western Avenue to the guard station near the Chicago River lock. No one ever claimed the derelict vessel and it was sent to the dump, along with the dozen steel drums that kept it afloat.

Another, among the elite for houseboats, had a name, street address, and was visited by a dignitary. Orlando W. Wilson, Superintendent of the Chicago Police Department, was brought to the city by Mayor Richard J. Daley to clean up the department and institute modern policing methods. He is remembered for changing the color of the rotating light on the roof of each police vehicle from red to blue, so as not to confuse police cars with fire engines. His wife, Mrs. O.W., made her own mark as chairman of the Chicago March of Dimes. To kick off the 1964 Mothers' March of Dimes door-to-door campaign on January 28, she rang the doorbell of Nicholas and Nettie Simmons on their houseboat, *Hi-Lo*, at 4114 North Rockwell. Nick, in his captain hat, and Nettie, in her head scarf to ward off the chill, greeted Mrs. O.W. on the gangplank in the early morning hours. The impressive scene was captured by a *Chicago Tribune* photographer—even the KEEP OUT/OFF sign and fish-cleaning table. Mrs. O.W. led 30,000 mother-volunteers, ringing door bells and collecting $250,000, one-third of the 1964 campaign goal, aided by members of the Illinois Loyal Order of Moose for security and use of police stations for campaign collection centers.

Houseboats eventually came under control by the District through its policing the discharge of waste and requiring holding tanks, pump-out capability, and a lease or permit. None could afford these requirements, or outfit the floating home, and eventually found other living arrangements.

CONTRACT CONDITIONS

It may have seemed like a gift to the District when the Northwest Land Association (Association) deeded the 13.7 acres in a 180-foot-wide strip of land for the North Branch Improvement in 1903, but the devil was in the details. (See Figure A-2.) The Association gave itself a bigger gift by conditioning the land transfer on having the District place the excavated spoil on the remainder of the land to fill in low areas and increase the usable land for development. But a few years later, the Association was not happy that construction of the new channel took longer than the agreed two years. After communicating their unhappiness to the District and claiming damages in 1909, they went to court. The lawsuit filed by the Association was just the first volley. Libby Hill describes this controversy on page 146 in her book, *The Chicago River: A Natural and Unnatural History*. In 1893, the Association purchased 400 acres bounded by Lawrence, Western, Montrose, and Kimball Avenues as an investment. The land was held in trust by Chicago Title and Trust Company. The North Branch wound its way through the land from north to south, normally being a thin thread of a meandering stream lined with trees and flanked by broad floodplains. In wet years and times of flood, the stream rose out of its banks to occupy up to 32 acres. Standing water and wet ground persisted long after the rains ended.

In 1903, 13.7 acres through the Association's property were deeded to the District for the construction of a new channel for the North Branch, which would confine the water in a wider and deeper prism, not only shrinking its expanse in times of flood, but providing a lower outlet for drainage of the adjoining land at all other times. The District optimistically agreed to construct the new channel in two years and spread all excavated spoil in the old river bed and adjoining low areas. The Association's investment was significantly enhanced, making more land available for development and draining the wet ground. A construction contract was executed by the District in early 1904 with Callahan Brothers & Katz Company from Omaha, Nebraska. Work began in May 1904 and stopped in October because Callahan didn't have the proper equipment to excavate the soft wet clay. It was intended to use conventional excavation equipment of the time, but the District plans didn't properly characterize the nature of the subsoil. Rather than litigate for breach of contract, the District agreed to an assignment of the contract to Great Lakes, the contractor for excavation south of Montrose.

Great Lakes was excavating by dredging, as this was more suitable for the character of the soft clay and less costly than dry excavation. A time delay was set in motion because Great Lakes couldn't start at Montrose until they had worked their way north from Belmont. Dredging at Montrose eventually began in June 1906, well past the two years agreed to with the Association for completion. Over a year later dredging was completed in August 1907. A small amount of the excavated material, the drier and stiffer clay, was excavated by dipper dredge and deposited in low areas south of Montrose. Most of the new channel was excavated by hydraulic dredging and the spoil was spread by sluicing north of Montrose. A large area including the old river bed and adjoining low areas were enclosed by a low dike. The soupy, clay-water slurry was sluiced into the enclosed area where the clay settled and the water either flowed back to the new channel or evaporated. For an extended time the enclosed area was wet, soft, and unattractive. It wouldn't be suitable for building houses until the wet clay consolidated and dried out, most likely two or three years. Several freeze-thaw cycles would help.

Chicago Title served notice to the District on behalf of the Association for damages, delay, and to recover the deeded 13.7 acres because the District didn't comply with the terms of the deed. Various claims were made and legal steps taken in 1909, and in 1910, a lawsuit was filed on these claims. The District counter-filed to enjoin the Association from acting on the claims. The trial court determined that the District didn't comply with the conditions in the deed, but due to the significant changes in the landscape, it was not possible to restore the land to its former condition. The court also found that the new channel was for a public purpose and its use could not be altered. On the question of damages, the court found that the Association was entitled to damages for the value of the land and the excavated material not deposited on Association land. Neither party was pleased with the outcome. The Association wanted more in damages and the District wanted none. The Association took the lead in seeking an appeal.

The decision in the appeal was not rendered until 1917 and, by that time, construction on the areas in dispute had already begun. For instance, Waters Elementary School at Campbell and Wilson was built and classes began in the fall of 1911. Obviously, enough population was in the neighborhood to require a school. Incidentally, the school was built in the old stream bed, but the name Waters didn't come from the North Branch. Rather, according to a plaque mounted in the school's first floor hallway, the school was named for Thomas J. Waters, chief engineer of the Chicago Public Schools, who devised a unique method of school ventilation. On the west side of the new channel, William E. Harmon & Company purchased 80 acres from the Association in 1908, platted the Ravenswood Manor subdivision, and installed alleys, sidewalks, streets, and utilities. Harmon built model homes and started selling lots in 1909. Numerous homes were built by 1917.

The appeals decision, long and detailed, cited a breadth of case law in upholding the trial court's clarifying the difference between a condition and covenant in the deed. The court also found that the Association was negligent in not bringing action when the conditions were violated, but waited until the new channel was completed. In fact, it was revealed that the Association had written letters to the District acknowledging the delay and requesting the work proceed expeditiously. In the 1909 letters claiming damages and requesting payment, the Association also acknowledged the difficulty in excavation. Further, in testimony, the District showed that the Association was not specific as to the areas to be filled and the quantities required for filling. Discretion was left to the District to have the fill reach a certain elevation, determined to be 11 feet above CCD. The appeals court reversed the trial court decree and ordered that the claims of the Association be dismissed and that a decree be rendered perpetually enjoining the prosecution of the claims. The Association's request for a rehearing was denied by the appeals court. This was the end of the litigation with the Association.

COVENANTS, DECKS, DOCKS, AND GARAGES

Perhaps before the court decision in 1917, but certainly not long thereafter, the first structure appeared on the channel bank within the 180-foot-wide right-of-way owned by the District. Homes and apartment buildings in the subdivisions on either side of the channel between Lawrence and Montrose were being built, and those adjoining the channel would be enhanced by a boat dock. After all, the District didn't erect a fence, didn't post signs, or even place obscure property line markers to define what they owned. The District, like an absentee owner, rarely checked to see what the neighbors were doing.

Along with other environmental awareness issues in the 1960s, the public began to complain about the appearance of the waterway. Environmental protection agencies at the federal and state levels hadn't been created yet and complaints were sent to congressmen or the City aldermen; the complaints invariably received the same response: *Call the Sanitary District.* The District could only send an inspector to verify the complaint; no program was in place to clean up floating debris or prevent the overflow of sewers. Those programs would eventually develop, but out of this agitation, fingers began to be pointed at boat owners using the channel and docking their boats. One finger pointed at the docks along the North Branch between Lawrence and Montrose, and the long simmering dispute of the right of occupancy.

In early May 1965, the District law department issued letters to about 20 homeowners ordering the removal in ten days of docks and other improvements located within the 180-foot right-of-way. For lack of compliance with the demand, the District threatened to remove the offending encroachment and seek reimbursement from the homeowner. The District claimed that its demand was for the purpose of maintaining the channel banks and to have a clear channel for flood control. The Ravenswood Manor Improvement Association (Manor) sent a letter to the District acknowledging the District's ownership and its purpose in sending the letters; and proposed that structurally sound docks be allowed to stay, the owners be given more time to comply, and a marina be built in Horner Park for a place to moor boats. The Manor president obtained a verbal commitment to extend the order to 30 days in a telephone conversation with Vincent Flood, the District's staff attorney who signed the letter. Perhaps it was a defensive measure when the Manor announced plans for a river bank cleanup the same month.

When challenged by a reporter as to why the District selected the homeowners with boat docks, given the many unsightly channel banks up and down the North Branch and the many houseboats moored north of Irving Park, Mr. Flood responded that "we had to start somewhere." The District attack on the homeowners became big news for a few days and even was the subject of a Mike Royko article on May 21, giving the famed columnist a chance to poke fun at Vincent Garrity, one of the more colorful District trustees. A month later, the District backed down and relented on their demand. The boat dock owners were allowed to keep their docks as long as they kept them in good condition and didn't allow debris to accumulate. The alleged problem faded.

Docks on the North Branch came back to the fore in the 1990s, again rising out of public concern about the river. Left to itself the District would allow these docks to be, but as is often the case when the public claims a public agency isn't doing its job, the public agency machinery overreacts. Prodded by the Friends of the Chicago River (Friends), the local alderman lodged a complaint with the District that private landowners were fencing off the channel bank and restricting public access to public property. Upon investigation, the District determined that 32 structures and 40 fences had been constructed within the 180-foot right-of-way. Fences and structures were separated to address fundamentally different issues that each represented, access vs. liability, respectively. The District held that a non-permitted structure exposed the District to liability if someone were to be injured or lose their life. The District also attempted to draw the City and Park District into the controversy, as the city had authority to issue permits for structures and the Park District could maintain a public path, if such were installed. After two meetings, the City and Park District bowed out, never to return.

To determine how significant the 40 fences and 32 structures were, let's put these in perspective. Due to the layout of the streets and lots along the 180-foot right-of-way, there were then, and are now, 16 adjoining property owners on the west side of the channel. (See Figure A-3.) The streets run east-west and the diagonal side lot lines adjoin the channel. All are single family homes except for two commercial buildings, one on Lawrence and the other on Montrose. On the east side, Virginia Avenue parallels the channel north of Sunnyside Avenue, and there are 39 owners that back up to the channel between Lawrence and Sunnyside. One commercial building at Lawrence and a small park at Leland Avenue also abut the channel. Single family houses dominate south of the CTA Brown Line and multi-family buildings dominate north of the Brown Line. Between Sunnyside and Montrose Avenues there are four owners with long diagonal side lot lines, one single family, two multi-family, and one parking lot. The District had a lot to deal with given the number of owners. Some multi-family buildings were condominiums, adding to the number.

The District staff sought guidance from the commissioners in October 1991, which turned out to be a request for public input. At a daytime meeting on a workday in January 1992, conducted by one of the commissioners, 20 people presented their ideas along with other commissioners expressing their viewpoints. All of the expressions were condensed to four suggestions: (1) have an evening meeting to allow more members of the public to participate, (2) take no action, (3) prosecute the trespassers and remove the fences and structures, and (4) legitimize the encroachments. The last suggestion was further defined in four ways: (4a) issue short-term permits to existing encroachments to protect the District while the owners obtained permits from the City, state, and Corps, (4b) establish a permit program for current and future uses of District property, (4c) convey the 180-foot right-of-way to the City, reserving an easement for flowage, and (4d) amend the District statute to allow negotiated sale of the channel bank to the owners.

The evening meeting was held in August at the Horner Park Field House and despite a big crowd, no further ideas were expressed to add to the suggestions categorized above. It was decided to work on the legislative proposal and the suggested authority became law in August 1994. However, the selling of public land was not supported by advocacy groups, principally the Friends, and a working group was formed to find a mutually agreeable course of action. The working group included the District, owners, advocacy groups, and the City, and they began meeting in November 1994. Meanwhile, the District agreed to stay the sale of property. The group met frequently, at least once per month; finding agreement was impossible and finding consensus was also difficult. There was agreement on a very general statement of consensus in October 1995, and the District agreed to continue the stay on the sale of property while the group attempted to elaborate the consensus statement. Meetings continued through 1995 and into 1996 without much progress. Sticking points were the required insurance, amount of detail required for a permit, permitting new structures, and permit oversight. Each of those would represent a significant cost to individual owners.

With little headway and a feeling that others in the working group besides the District and owners had little to contribute, the owners gave birth to their own organization, the Ravenswood River Neighbors Association (Neighbors) in early 1997 and made a very specific proposal. The Neighbors would become the District partner and assume the cost of insurance, permitting, and oversight, suggesting that the District would only have to deal with the Neighbors rather than 67 or so owners. The Neighbors had already taken the initiative to have the Illinois Department of Natural Resources inspect the channel and improvements and issue two general permits for residential piers and shoreline protection.

This might have done the trick, except for an incident in the March 1997 flood: a boat owner complained to the District of high water and requested the water level be lowered so that his boat tethered to the dock wouldn't capsize. It was found that the boat owner was renting the dock from the adjoining owner. While this improper act might have been remedied by the oversight responsibility of the Neighbors, the renting of the dock galvanized feelings at the District that the existence of docks was fundamentally wrong; adjoining owners were making private gain on public property.

In April 1997, the District staff suggested the following courses of action to the commissioners:

1. prohibit and evict current occupants, and refuse future requests for occupancy;
2. initiate the sale of channel bank sections to the adjoining owners;

3. issue a blanket permit as suggested by the Neighbors; and
4. issue individual permits or a blanket permit on an interim basis, requiring insurance and eventual removal of structures.

The commissioners punted and decided instead to hold another public hearing, which occurred in May, again in the evening at Horner Park. Nearly 100 attended the meeting and 11 persons offered comments in addition to the two aldermen and commissioners. While the District staff advocated for the fourth course above, the consensus of the meeting was between the third and fourth courses. In June, the District issued a notice laying out the criteria for issuing and administering five-year permits and allowing the Neighbors to carry the insurance for all owners. A draft form of a permit and the criteria were discussed with the Neighbors and both were unacceptable.

Despite further discussion between the District and Neighbors, the owners couldn't stop feeling that they were being treated like trespassers and irresponsible public stewards. The District, while acknowledging that many of the docks and channel bank improvements were pleasant to the eye and structurally sound, seemed to only focus on the few dilapidated and unsightly docks. The District also kept referring to its liability exposure and its undisputed ownership. On the other hand, the Neighbors pointed out that no one was filing personal injury lawsuits and the owners did have a right of access via the 1903 covenant. The District repeatedly held that the covenant had no validity. The District staff issued a slightly revised set of permitting criteria, which was thought to address the concerns of the Neighbors, but which still referred to the dock owners as *encroachers*. Little came of this latest effort to reach agreement.

As things happen with a stalemate, the issue drifted for a few years. The District used the time to survey the 180-foot right-of-way, catalog the encroachments, and identify the encroachers by name and address. Finally, in July 2003, the District struck, filing lawsuits against the encroachers to forcibly evict them from the premises. The Abec case was so named because Abec was the first name in the alphabetic order of all owners to be sued (and all individual lawsuits were consolidated). The action put the matter squarely in the public eye and articles appeared regularly in various newspapers, with occasional footage and sound bites on broadcast news.

The Friends cheered the District for invoking the public trust and protecting public land. Just the same, amnesty would be extended during the pendency of the lawsuit if only the encroachers would comply with the permit criteria and apply for a permit. However, the cost of the permit was undecided. Some suggested a modest $200 per year and others wanted up to $1,000. The District began to appraise the property, suggesting that the permit fee be set as a percentage of market value or be based on the lineal feet of water frontage. Also upping the encroachers' costs was a new requirement, to have a licensed structural engineer certify the integrity of each boat dock, deck, or patio.

The collapse of a few backyard porches in the Lincoln Park neighborhood, resulting in some deaths, emboldened the District, which declared the porches emblematic of what could happen to a boat dock. The improvement in water quality, brought about by Clean Water Act requirements, was cited as the reason why so many people wanted to boat on or walk along the channel. It was also suggested that the boat docks and decks added to the value of the owners' property, *so why shouldn't the permit fee be indexed to the property value?* To the Neighbors, all this posturing and pontificating by the District was irritating. Since both parties were at odds over the validity of the 1903 covenant, why not let the court decide its validity? So the Neighbors filed a lawsuit seeking summary judgment on the last business day of 2003. Adding insult to injury, three homeowners, two west of the channel and one east, whose garages were encroaching on the 180-foot right-of-way, received notices on Christmas Eve 2003 that their garages, or parts of them, had to be torn down.

The court took its time to consider the 1903 covenant and removal of structures; in September 2006, the court found in favor of the District—that it was the rightful owner of the 180-foot right-of-way, subject to the conditions in the 1903 covenant. For the Neighbors, the court found that the owners have an easement right of ingress and egress to the water's edge until the District widens the channel, but the easement right extends only to a dock, but not to the impermissible structures on District property. It also clarified that the District permit fee is not a special assessment for widening the channel, and the District may not enter and remove the structures until further court proceedings determine that removal is necessary in providing for a water channel. The covenant lives!

A half dozen years later, in November 2012, the court approved a consent agreement between the parties on the remaining issues, settling for all time the opposing views on the covenant. The agreed order merged the terms and conditions of the 1903 deed into the agreed order as they pertain to the District and the Association's successors. The 1903 covenant died and was reborn as authority attached to the deed for each adjoining owner, preserving the essence of the covenant in a new form. The agreed order also includes the legal description of the 180-foot right-of-way in the deed for each adjoining owner. The District was directed to have the agreed order recorded on Cook County records, giving it permanency for all current and future property owners.

On other disputed issues, the agreed order clearly states that:

- The District has the right to reasonably regulate their property and have reasonable right of access to the owners' easements.
- The District may assess an annual administration fee of $300 for each owner occupying the channel bank; and the Neighbors shall obtain and maintain insurance for the owners occupying the channel bank as required by the District.
- Each owner must submit an inventory of existing fences and structures within 120 days and may continue to use the existing improvements as *authorized structures* so long as they are kept in good repair and the owner is in compliance with District requirements.
- The District may remove improvements not included in an inventory, or when they become in disrepair or act as obstructions to flow in the channel.
- Future or replacement structures are only allowed by permit issued by the District;
- Any encroaching fences or structures 2.5 feet or less into the 180-foot right-of-way are considered *de minimis* and included as *authorized structures*.
- The three garages that are more than 2.5 feet into the 180-foot right-of-way are not authorized, and the three owners shall each pay a prescribed annual rental.

The agreed order further provided that both the District and Neighbors will each dismiss their lawsuits and claims against the other, that each party will pay their own legal expenses, and that to resolve future disputes, both parties shall engage in a dispute resolution process. If the dispute resolution fails, the dispute may be brought back to the court. While the agreed settlement under the authority of the court appears to have resolved all issues in dispute, it remains to be seen if harmony will prevail in perpetuity. Yet all is well so far: the property owners have submitted their inventories, the necessary fees and rentals are up-to-date, and no applications for new structures have been submitted.

Appendix A-6: The Panama Canal and Mississippi River Commission

The trustees decided in January 1914 that a commission was needed to study the problem of sloughing channel banks for both the Calumet-Sag Channel and North Shore Channel. Two engineers from the staff and three trustees were designated and their report was issued in April. Dubbed *The Panama Canal and Mississippi River Commission*, they departed from the city in early February. In Memphis, they learned from the Corps about the use of weighted fiber mats and slope paving along the Mississippi River for slope protection. However, this method was determined as not applicable for the District as it didn't protect against the freeze-thaw cycle or large fluctuations in water level. In New Orleans, a seawall on Lake Pontchartrain and the City's water and sewer systems were inspected. It sounded familiar when they learned that mortality had declined after sewage was collected and piped away from the city. The Galveston/Houston area was next, but like New Orleans, nothing observed pertaining to slope stability was applicable to the District.

The group arrived by boat in Colon, Panama, in mid-February and traveled to Panama City by rail. Following an orientation, the Culebra Cut was visited, inspections conducted by foot and boat, and interviews held with the Panama Canal engineering staff. Part of the Culebra Cut excavation was in clays similar to conditions back at the District. Although the size and complexity of the undertaking in Panama were greater than in the District, there were some similarities. A majority of excavation was in dry conditions and the cut was flooded to retard the rate of sloughing. Maintenance dredging in Panama was used to provide channel capacity for navigation and to eventually attain the final channel depth and width. Disposal of dry excavation and dredged spoil was usually far from the canal route within the ten-mile-wide Canal Zone.

The commission focused on conditions in the Calumet-Sag Channel and North Shore Channel, comparing the two. The former had shallow layers of clay and peat soils deposited on top of bedrock, whereas the latter had thicker and deeper layers of clay soils above bedrock. In both, most slope failures were the result of placing spoil atop the channel bank, increasing the load upon the native soils and overtaxing their bearing capacity. To a lesser extent, slope failures were the result the lateral movement of moist soft clay when exposed by an open cut. A third cause of slope failure would be found once the channels were placed in operation, where erosion occurs when subjected to the erosive force of moving water caused by floods or passing boats. Comparisons were attempted between conditions in the Panama Canal and conditions in both channels back in the District, but since much of the Culebra Cut was through rock formations with active fault zones, the extent of comparisons was limited.

The commission prepared a report on these finding and included extensive discussion of the conditions along and construction of both channels in the District, embellished with numerous figures and photographs, and references to the experience of constructing the Sanitary & Ship Canal. The report concluded with two glaring distinctions between the District channels and the Panama Canal: (1) the District relies on local taxes limited by state statute to fund the work, whereas in Panama the executive branch was given carte blanche by the legislative branch for a successful outcome; and (2) the District had to acquire, often by condemnation, all necessary right-of-way for the construction of channels, whereas, a ten-mile-wide strip was made available by treaty to the U.S. government for the Panama Canal. With this caveat, the report concluded by finding the District using the best engineering methods available to meet its goals within the limitations imposed by federal and state government.

Appendix A-7: Skokie Dream Doesn't Become a Nightmare

Mayor Al Smith of Skokie thought it was an exciting concept, a 400-boat marina and motel on the eastern edge of the landlocked village. In October 1982, this idea had been given momentum by District planning to change the North Shore Channel. Deep tunnel had been under construction since 1975 and it was three years until the tunnel under the channel would be placed in service, but it was believed that Deep Tunnel would be so immediately successful that the channel would meet the highest state water quality standards. The four-mile reach of channel from Sheridan Road to Oakton Street could be remade as an estuary off Lake Michigan by removing the lock and pumping station under Sheridan Road and building a new lock and pumping station at Oakton Street.

When the tunnel went into operation in 1985, it reduced the quantity of combined sewer overflows from outfalls owned by the District, Evanston, Skokie, and Wilmette, but didn't eliminate them. To this day, water quality in the channel is occasionally compromised by combined sewer overflows following large rainstorms.

Historically, the District didn't encourage boat traffic on the channel because it was narrow and the banks were steep and unstable. Even though the lock gates were long gone, the large sluice gate at Sheridan Road was designed to lift high enough for boats to pass under it. If the marina project were to go forward, it was recognized that the channel would have to be widened and the fluctuating lake level might require bridges to be raised. Skokie requested the Corps to make a feasibility study. Two years later the Corps had not finished the feasibility study and the District began its own study, its interest being in enhancing the leasing value of the land it owned along the channel. The District had also been considering a land use plan for its properties along the Calumet-Sag Channel.

However, broad community support for the Skokie marina plan was lacking. Evanston and Wilmette leased District land for a golf course and an arboretum and opposed the marina plan as it would set a precedent for commercial development along the channel. Citizens proposed that existing unleased land be cleaned up and made available for public use, using the plan for the Calumet-Sag Channel as an example. By 1987, the commissioners and community officials from municipalities along the channel had come together to endorse the concept of public use and access to the channel water edge. A set of guidelines resulted—the River Edge Renaissance—and they were applied to virtually all District riparian property. Cooperation of the District and several local agencies resulted in many park developments along the channel and a continuous biking/hiking path from Green Bay Road to Lawrence Avenue.

Another plan of the District for flood control launched in the late 1950s would have seriously changed the North Shore Channel and the Wilmette Pumping Station. The plan was based on expediting the removal of floodwater and included a pumping capacity at Wilmette of 9,000 cfs to discharge floodwater to Lake Michigan and a channel capacity of 5,000 cfs, more than doubling its size. Fortunately, the plan was aborted in the early 1960s following the appointment of Vinton Bacon as General Superintendent, who advocated for temporarily storing stormwater before releasing it downstream. Mr. Bacon was aware of proposals for federal water quality legislation to require strict standards that would reduce, not increase, the discharge of combined sewer overflows. The federal Water Quality Act became effective in 1965.

Appendix A-8: North Shore Channel Land Opportunities

REAL ESTATE OPPORTUNITIES

Initially, the District had no plan to use or maintain the land along the channel. An inquiry from the River Park District opened the possibility of establishing a roadway for access, and the park District was granted a permit in August 1916 to use the land from Lawrence to Devon Avenues for park purposes. The Chicago Park District, successor to the River Park District, was granted a lease and continues to use the land at Legion, River, and Ronan Parks.

The Essanay Film Manufacturing Company leased the area along the east side of the channel between Foster and Bryn Mawr Avenues for a period of three years beginning in October 1915. Essanay was not allowed to build any permanent structures, could use the site only for making moving pictures or photographs, could not have liquor on the site, had to pay for all utilities and taxes, could not create a nuisance or use the site for illegal or immoral purposes, and many other restrictive limitations. The District apparently felt that filmmaking needed moral guidance. These restrictions may have been why no movies are known to have been filmed on the site. However, the spoil banks must have appealed to the Essanay moguls until they took their business to the real Wild West.

The National Rifle Association requested target space along the channel, but the trustees denied the request. The District issued a permit in September 1926 to an entrepreneur for use of land near Foster Avenue for an airport to facilitate the city being a hub for the national airplane service in the transport and delivery of airmail. An airport was never developed.

McCORMICK BOULEVARD

Not known for road building, the District launched a big plan in April 1923 to construct roads along the three built channels—another attempt to fulfill the decades-old dream of a commercial/industrial corridor along the waterways. The reasons:

- Allow access to waterfront property, thereby enhancing the value of leasing over 4,000 acres of potential dock property.
- Expedite the removal of spoil encumbering this property.
- Connect suburbs to the city and relieve congested streets.

The last reason reflected the poor condition of roads outside the city; few roads were paved and most were rutted, poorly drained, and not well maintained.

The District lacked authority for road building, but that was soon remedied by an act of the General Assembly the same year. The first road planning focused on the Sanitary & Ship Canal and North Shore Channel, and the latter offered least resistance to construction (fewer railroads and major thoroughfares to cross, less spoil to be removed, and freer access through less densely developed areas). However, not all the right-of-way of the channel was easily accessible for a roadway. South of Devon Avenue, it would involve the City's approval of plans and construction details and numerous permits, as well as disrupting the lease with the Chicago Park District. From Belmont to Lawrence Avenues, residential properties would have to be acquired due to the limited 180-foot right-of-way for the North Branch Channel. In Evanston and Wilmette, east of West Railroad Avenue, there were two railroads to cross and potential opposition due to proximity of residential neighborhoods and recreational land use. Clearly, the first road building project would be the 4.5-mile segment from Devon to West Railroad Avenue.

While detailed design proceeded, the first contract for remaining spoil removal and rough grading was awarded to the L.E. Myers Company in April 1924. The segment was divided into two equal sections for construction, and contracts were awarded in July and August to the John A. McGarry Company and the James A. Sackley Company, respectively, for a 40-foot-wide roadway with all necessary appurtenances for drainage and intersections with street crossings. An agreement with the Chicago & North Western Railroad in April 1925 provided for building a viaduct for the Mayfair Division south of Oakton Street with an underpass for the roadway. In October, the M.E. White Company was awarded a contract for construction of the viaduct, approaches, and the included section of roadway. The railroad built the diversion for its tracks, new roadbed, and tracks, the cost thereof to be reimbursed by the District. Named in honor of Robert R. McCormick, president of the District from 1905 to 1910, the road was opened in 1926. An engraved granite monument can be found northeast of the intersection of Devon Avenue and McCormick Boulevard.

Eventually, the right-of-way of McCormick Boulevard was transferred to the state of Illinois. The Illinois Department of Transportation has modernized the roadway more than once and removed the railroad viaduct south of Oakton Street.

MILLION DOLLAR BRIDLE PATH

It was either while riding in the backseat of our 1941 Plymouth on the way to the drive-in movie theatre on McCormick Boulevard at Touhy

Avenue, or perhaps on an outing in the north suburbs, that I remember dad or mom saying something about the *million dollar bridle path.* I'd look away from my Superman comic and gaze out the window, but I never saw any horseback riders. This memory came back to me while doing research, but I never saw a reference to the bridle path in the board proceedings and there was good reason why. That stretch of land between McCormick Boulevard and the canal became an attractive opportunity for greedy politicians and bureaucrats, turning out to be one of a few scandals that bedeviled the District. It happened in the late-1920s and culminated in the early-1930s, sometimes referred to as the *whoopee era*, with some being sentenced to the penitentiary for conspiracy.

The District was seeking legislative authority for a bond issue in 1927, and somewhere along the way someone must have asked "Where's mine?" In order to please, a scheme was hatched, employment ballooned, and some contractors not under contract received payments. Civic watchdogs raised questions and put pressure on the trustees. The 1928 election only brought two new trustees to the board, but the incumbent president lost his bid for reelection and was on the way out. With a new president in December, the employment roll began to thin dramatically, and by the end of the year nearly 1,000 were given pink slips, most from the law and plant departments. It was no surprise that the Citizen's Association, which helped to create the District, demanded that the Cook County State's Attorney (State's Attorney) initiate a grand jury investigation. By coincidence, the State's Attorney was the brother of the ousted president.

The Illinois Senate investigating committee was frustrated in February 1929, because the State's Attorney wouldn't release the financial records of the District. It was claimed that release of the records would interfere with the investigation being conducted by the grand jury into corruption at the District. But the senate committee was on to something. A similar request for records directly to the District received the same response. These were not the public records, but the books, check registers, petty cash, expense accounts, etc. The news media were hinting at contracting irregularities and ghost payrollers.

A bond issue authorized in 1927 had been invalidated by a court, and the urgency of the Senate committee's request was tied to new legislation being expedited through the legislative process again authorizing a

$27 million bond issue for the District. With rumors of scandal, the committee wanted assurance that the money would be honestly spent on justifiable projects. The committee intended to advise the governor before he was asked to approve the legislative action.

The U.S. District Attorney and State's Attorney each initiated grand jury investigations, the former for tax evasion and the latter for conspiracy to commit fraud and misuse of public funds. Several contractors were indicted by the federal government early in the 1930s. By the mid-1930s the state indicted six District trustees, two department heads, and nine employees. About 1,500 witnesses were called before the grand jury in the state's case. The federal case netted six contractors and some contractor employees and also found salaries paid to ghost employees, who were believed to include some of the trustees.

Details in testimony in the county circuit court hearings were outrageous. To avoid the statutory contracting requirements, contractors were being paid from petty cash funds in checks under $500; invoices were falsified and District employees were knowingly approving work for removing nonexistent spoil and placing cinders on the bridle path, the same cinders the contractors were paid to remove from District plants and pumping stations. Contracts were awarded to nonexistent companies and false invoices were submitted for imaginary work. So-called "District employees," several hundred in number, admitted receiving paychecks and not putting in time at any District work site over a period of a year and a half. Some were even submitting false expense reports and receiving payment for expenses not incurred.

Many of the so-called employees were state senators or representatives, relatives of members of the legislature, or campaign workers. Some of the nonexistent companies were fronts for members of the legislature. One group of downstate attorneys, known as the Beardstown buccaneers, were paid to inspect and process bogus flood damage claims, but they didn't set foot along the Illinois River or submit any reports. Trial records revealed that about 40 members of the General Assembly were on the District's payroll and another 60 members sponsored others who were on the payroll but did no work.

However, the State's Attorney didn't bring charges against the companies, contractors, or so-called employees, apparently because of their willing testimony. Rather, the trustees and department heads were being tried for conspiracy in creating fraudulent schemes and misuse of public funds. One trustee was acquitted since the conspiracy was created during 1927, the first year of his first term; evidence was lacking linking him to the conspiracy. Another trustee died during the trial. Two other trustees were acquitted because the evidence could not establish their guilt beyond a reasonable doubt. Two trustees, one the former president and lead conspirator, and two department heads were found guilty and sentenced from one to five years in the state penitentiary. The former president died of a heart attack while his sentence was being appealed.

It was claimed that the bond issue was necessary to build treatment plants and sewers required to clean up the canal. The bridle path became notorious as a frivolous expense, but it was claimed that the public wanted the land between McCormick Boulevard and the canal to be cleaned up and made available for recreation. An equestrian path for the North Shore seemed appropriate.

The indicted opted for a bench trial to avoid the wrath of a jury composed of tax paying citizens. However, the former president even attempted to walk out of the courtroom when the prosecutor was telling the judges of his misdeeds. All three county judges on the trial panel were astounded at the brazen conduct and greed of the elected public officials—those on trial and those who went along with their votes on legislation or contracts. However, the judges could act only on the charges and evidence. Although there was speculation in the news media that some of the misused public money made its way back into the coffers of both political parties, no charges of such acts were brought or evidence presented.

The judges expressed regret in having to find elected public officials guilty as charged, as it created mistrust of elected officials in the minds of the public. Interesting tidbits from testimony: one person receiving a District paycheck stated that his work was coaching the former president's baseball team; a downtown hotel barber was on the payroll and only cut the hair of politicians; when asked how the trustees acted when he read the list of fraudulent expenditures at a board meeting, the clerk of the board indicated they were cracking jokes and puffing on their cigars; a limousine company was richly rewarded for trustee in-town and out-of-town travel, including female companionship, food, and liquor; a secretary in a travel agency was on the payroll to arrange personal travel for trustees; the secretary's husband, owner of the travel agency, arranged for a District-paid trip to New York City for several trustees that included lodging at the old Waldorf-Astoria Hotel. It came to light that a party at the hotel became unruly, resulting in several thousand dollars damage to hotel furniture. The hotel's claim was paid, but the source of the funds was not disclosed.

Another inquiry was launched in November 1928 by the Chicago Bar Association (Bar Association) regarding the conduct of its members swept up in the scandal, which resulted in disciplinary action recommended to the Illinois Supreme Court in June 1930 for 55 attorneys. Most prominent among the 55 was the head of the District's law department, grandson of a Bar Association founder, and son of a highly-regarded former State's Attorney. The result: only one attorney was disbarred, 21 were censured for dereliction of professional duty, one was suspended from practicing law for two years, and 13 were suspended for 90 days. The head attorney, one of the 90-day suspensions, claimed that he was following orders from the president. The court admonished him for placing duty to his client above his duty to the public and ethical conduct. The remaining 17 were exonerated because they were able to show that they faithfully performed all duties assigned for the time paid, even though they were given far less than full-time work.

The head attorney did have a point, as back in those days the president appointed all employees, and the trustees decided the award of contracts limited only by the requirement for competitive bidding. Eventually, the General Assembly wisely removed the authority for administration, contracting, employment, and day-to-day operations from the duties of the elected trustees.

In February 1932, the District initiated civil suits to recover the ill-gotten gain against nearly 1,000 former employees, contractors and the persons behind the nonexistent companies who received fraudulent payments. However, the money was long gone and the nation was in the depths of the Great Depression; there was little likelihood the money would be recovered.

Appendix A-9: O'Brien Water Reclamation Plant

This regional water reclamation plant reclaims water from sewage for a large part of northeast Cook County extending from the Cook-Lake County border on the north, Lake Michigan on the east, Fullerton Avenue on the south, and Harlem Avenue and the Des Plaines River on the west, an area of 143 square miles. The plant serves a population of 1.3 million and processes 230 million gallons of sewage on an average day. During wet weather, the plant can handle 500 million gallons per day.

Reclaiming water from sewage involves biological and physical processes. The physical processes are mechanical screening and gravity settling; the biological process is the enrichment of microorganisms in sewage with oxygen to digest organic matter in the sewage. The microorganisms derive from the human excrement in the influent sewage. No biological agents or chemicals are routinely used for treatment. Reclaiming water from sewage mimics the natural process by which a river cleanses itself. However, the process is engineered to do in a few hours what nature takes days to accomplish.

Three intercepting sewer networks totaling 99 miles in length convey sewage by gravity to the plant from the many cities, villages, and unincorporated communities in the service area. Three large intercepting sewers, the largest being 15 feet in height, all converge at the intersection of Howard Street and McCormick Boulevard, 40 feet underground. In the description of various processes below, the approximate travel time of liquid through the process is given. The actual time varies depending on operating conditions.

PUMP AND BLOWER BUILDING

Before being pumped up to ground level through one of six large electrically driven 1,000-horsepower pumps, the sewage passes through coarse bar screens with 3.5-inch openings to remove large objects that could damage the pumps (process: *physical separation*). The screenings are collected and placed in dumpsters. This building also contains seven electrically-driven blowers as large as 2,500-horsepower each to deliver air for the aeration process. Aeration is explained below.

FINE SCREEN AND GRIT BUILDING

After being pumped up to ground level, the sewage flows through an underground conduit to this building, where the sewage passes through fine bar screens with 0.75-inch openings to remove small objects. The screenings are collected and placed in dumpsters. The sewage passes through several parallel tanks where grit settles to the tank bottom (process: *gravity separation*). It takes about 15 minutes for the sewage to flow through these tanks. The grit is raked to a sump and piped into the dumpsters along with the screenings.

PRIMARY SETTLING TANKS

Next, the sewage flows outside through one of 16 large primary settling tanks. It takes about an hour for sewage to flow through each tank. Scum, collected by skimming the surface, is piped to concentration tanks to reduce the water content (process: *physical separation*). The scum is deposited in dumpsters and the contents of these dumpsters and the dumpsters described above are disposed at landfills along with municipal garbage. Solids settle to the bottom of the primary tanks and are raked to a sump from which the sludge is piped to sludge concentration tanks.

AERATION TANKS

Having passed through the primary process and then conveyed to the open aeration tanks, the sewage enters the activated sludge process, also known as secondary treatment. In the aeration tank, the sewage is mixed with large amounts of air piped from the blowers mentioned above; the air is diffused through porous plates in the side of the bottom of each tank. The rising bubbles produce a spiral flow that maximizes mixing and diffusion of oxygen into the sewage. Thorough mixing brings the organic waste in the sewage into contact with the microorganisms, often referred to as bugs. The microorganisms thrive in the presence of oxygen and digest the dissolved and suspend organic matter in the sewage (process: *biological digestion*).

There are 88 aeration tanks. Original tanks, 72 in number, are each 440 feet long. The balance of 16 tanks are shorter. It takes about six hours for the mixture of microorganisms, oxygen, and sewage to flow through each tank. By the end of this time, the microorganisms have grown and become linked together in a large fluffy mass called *floc*.

FINAL SETTLING TANKS

In the quiescent final settling or clarifier tank, the floc/water mixture enters at the bottom center, and as it flows upward toward the surface, it disperses radially toward the outer perimeter of the tank, travelling progressively more slowly so that the floc can gently settle to the bottom. By the outer edge of the tank, clear water leaves the top of the tank passing over the effluent weirs. The effluent, with more than 95% of the contaminants removed, is reclaimed water discharged to the North Shore Channel. The effluent meets all limitations and standards imposed in the permit issued by the Illinois Environmental Protection Agency.

There are 64 final clarifier tanks, some square and others circular. It takes about two hours to flow through the final clarifier tank to reach the effluent weir. The settled floc on the bottom of the tank, called activated sludge, is moved to a sump in the center by a rotating rake and is pumped to the head of the aeration tank as seed or starter for the activated sludge process. A small portion of the activated sludge is diverted to the sludge concentration tanks.

SLUDGE CONCENTRATION TANK

Sludge from the primary and final settling tanks is held for about six hours to reduce the water content. Water drawn off is drained to the plant influent to mix with influent sewage and go through the entire treatment process. A portion of the sludge from the concentration tank is drawn off and pumped through a pipeline 16 miles in length to the Stickney Water Reclamation Plant for further processing into a soil-like product used as a fertilizer or soil amendment called biosolids.

DISINFECTION

Recently placed in operation is an added process step that radiates the final effluent with ultraviolet light to kill bacteria and other microorganisms to meet a new required plant effluent limit and water quality standard for the North Shore Channel. This process was placed in service March 2016. Ultraviolet radiation is the same as sunshine, but engineered to do in seconds what takes hours on a sunny day.

Disinfection inactivates (kills) microorganisms in the effluent, commonly called bacteria. Both good and bad bacteria, including some pathogens, are inactivated. The regulated effluent limit and water quality standard is based on the quantity of fecal coliform colonies per 100 milliliters of water forming on a media plate examined under a microscope. Fecal coliforms are microorganisms from the digestive tract of humans and other warm-blooded animals; those microorganisms are used as an indicator of the total bacterial count because it is a common, inexpensive, and simple analysis to perform. Fecal coliforms do not cause disease but may indicate the presence of pathogens. The fecal coliform water quality standard is based on criteria developed by the U.S. Environmental Protection Agency from studies at beaches. The criterion is a statistical determination based on a risk of illness of eight or more illnesses per 1,000 persons ingesting or immersing in water. Below this incidence of illness, the water is considered safe.

OPERATING PERMIT

The plant operates and discharges effluent under a National Pollution Discharge Elimination System (NPDES) permit issued by the Illinois Environmental Protection Agency. The permit sets limits on the discharge and requires monitoring and reporting to verify compliance. Monitoring is performed every day and reports are submitted monthly. The agency also conducts periodic plant inspections to verify compliance. Permit details can be found on the District's web site.

Appendix A-10: Ellis Sylvester Chesbrough

Chesbrough was born in Baltimore County, MD, on July 6, 1813, received minimal formal education, and was put to work at an early age to help support the family. At the age of 15 he began working with his engineer father on construction of the Baltimore & Ohio Railroad, learning by experience the practical aspects of civil engineering from military engineers. The Panic of 1837 cooled railroad expansion and Chesbrough and his father turned to farming in western New York state, only to fail. Fortunately, he found work in water resources, working under the direction of John B. Jervis on the location and construction of aqueducts and reservoirs for the Boston Water Works. He accepted an appointment at age 33 as chief engineer of the western division water works, responsible for the Brookline reservoir project. Four years later he was made the commissioner of the water works and the following year, 1851, became city engineer of Boston. Chesbrough researched and published a report on sewerage practices in England, raising his profile in this new and challenging field.

On October 1, 1855, Chesbrough was appointed chief engineer of the Chicago Board of Sewerage Commissioners, a board whose creation was authorized by the Illinois General Assembly earlier that year. He immediately went to work and by the end of the year produced the first plan for sewers; construction began in 1856. Over the 1856–57 winter Chesbrough travelled to Europe on assignment by the commissioners to study sewerage systems in large cities. As the Chicago River and its branches suffered from the discharge of sewers, Chesbrough suggested a large canal to reverse the flow of river water into the lake. His suggestion went unheeded for three decades.

As a result of his experience in Boston, Chesbrough was also helpful to the Board of Water Commissioners on matters affecting the Chicago water works. In 1861, the two boards for sewers and water were combined into the Board of Public Works, and Chesbrough was appointed chief engineer. He conceived the idea and oversaw construction of two large projects: the two-mile water tunnel and intake crib out in the lake, and the pumping station and water tower on Michigan Avenue, a project that gained international attention and was displayed in France at the April 1867 Paris Fair. Following the Chicago Fire in 1871, another water tunnel was constructed six miles from the intake to a new pumping station on the West Side.

Chesbrough was appointed to the new position of Commissioner of Public Works, but his tenure lasted only four months; he resigned in 1879 to enter the practice of consulting engineering, working on several sewer and water projects in large U.S. cities and water resource projects in France and Spain. He was president of the Western Society of Engineers in the city in 1869 and president of the American Society of Civil Engineers in 1877. Chesbrough died on August 18, 1886. Had he lived a few more years, he would have witnessed realization of the big canal he had recommended many years earlier.

Appendix A-11: Isham Randolph

Perhaps the District trustees thought that permanency in the chief engineer position was elusive; between June 1890 and June 1893, four chief engineers had been appointed and four had resigned. Isham Randolph proved otherwise; he stayed 14 years and oversaw the completion of many projects: the 28-mile Sanitary & Ship Canal; 15 miles of Des Plaines River diversion channels; the five-mile Des Plaines River improvement in Joliet; the five-mile improvement of the South Branch; the four-mile extension of the Sanitary & Ship Canal, including the hydroelectric powerhouse, navigation lock, and butterfly dam; the 2.2-mile North Branch improvement; construction of 36 bridges; and replacement of 14 bridges over the Chicago River and South Branch. In addition, Randolph oversaw planning and design for projects constructed after his departure including the eight-mile-long North Shore Channel and 16-mile Calumet-Sag Channel.

Randolph, born in New Market, Virginia, in March 1848, attended private schools and worked on the family farm. Following the Civil War he began employment in field work for eastern railroads, learning surveying and construction skills. He supervised construction of parts of the Baltimore & Ohio Railroad system and freight houses and terminals for the Chicago & Western Indiana Belt Railway, eventually becoming chief engineer for the latter company in the city. He opened his own practice in 1885, continuing to serve railroads as well as other clients, including the Union Stock Yards. His extensive experience, self-taught engineering skills, common sense, and honesty impressed the District trustees and led to his appointment in June 1893.

In June 1905, Randolph was appointed to the Board of Consulting Engineers for the Panama Canal Commission by President Roosevelt; he was granted a leave of absence to attend to this duty and was away from September through November. Randolph's work for the Panama commission was critical to the future of the project; he was one of five engineers, a minority of the board, which recommended the high-level canal across the isthmus with locks at each end. The other eight members recommended a sea-level canal. The Panama Canal Commission and President Roosevelt adopted the minority recommendation. Randolph returned to the city in early December in time for the new board to take office following the November 1905 election.

The new District board was not only a change in political party affiliation, but its president was a scion of the McCormick family. A descendant of Joseph Medill, founder of the *Chicago Tribune*, McCormick liked to run things his way and perhaps led to Randolph's decision to resign in July 1907. The parting was not hostile—that was not Randolph's way—and rather than let go of a valuable engineer the trustees appointed Randolph as consulting engineer. He continued to serve the District with his engineering talent.

McCormick presented a statement citing Randolph's 14 years of service in accomplishing engineering work of great size and complexity, much of which was beyond the comprehension of lay minds and the understanding of the trustees. Randolph's service was without a cloud of suspicion and he served boards of different political orientations faithfully and without partiality. McCormick expressed his pleasure in being associated with Randolph and was grateful for his willingness to continue his service to the District. McCormick closed his remarks with these words: "The public is indeed fortunate in securing the services of so educated an engineer, so scrupulous a man, and with all, such a courteous gentleman." The trustees then adopted a resolution praising Randolph for his years of service, dedication to the cause of the District, preeminent abilities, and qualities of mind and heart. The resolution closed with these enduring words: "...and this resolution be entered in the proceedings of this board and that a copy, suitably engrossed, be prepared and presented to..."

Randolph continued to serve not only the District but the public as well, travelling again to Panama to witness canal construction progress, designing canals and harbors in south Florida, planning a harbor for Milwaukee and transportation and water projects in Toronto, advising on the construction of dams and locks for the Illinois Waterway, and consulting on flood protection along the Ohio River in southern Illinois. Isham Randolph's son, Robert Isham Randolph, followed in his father's footsteps as an engineer and civic leader in the city. Isham Randolph died in 1920.

Appendix A-12: Robert R. McCormick

Robert Rutherford McCormick was known as *The Colonel* later in life, but as *Bert* in his young adulthood.

Bert, born in Chicago in 1880, was the grandson of Joseph Medill, founder of the *Chicago Tribune*, self-proclaimed as the World's Greatest Newspaper. His father, Robert S. McCormick, was ambassador to Russia before President Theodore Roosevelt transferred him to France in 1904. Robert was not particularly respectful of royalty; hence he didn't get along with the Romanovs. Bert's mother, Katherine (Kate) McCormick, was fond of Medill, her oldest son, but not of Bert. She was domineering and distant, as she preferred living in Europe rather than in the city. Brother Medill suffered from nervous disorders, and in the fall of 1908 Kate sent him to Europe to recover. Later, he was placed in charge of a newspaper in Cleveland to prepare for taking over the *Tribune*. The Cleveland newspaper suffered from his mismanagement.

Elinor (Nellie), sister of Kate, was the wife of Robert Patterson and the mother of Joe Patterson. Kate and Nellie were often at odds and the cousins, Bert and Joe, could manipulate the two sisters for their own benefit. Robert Patterson became the editor of the *Chicago Tribune* after Joseph Medill's death. Kate wanted Medill to succeed Robert at the *Chicago Tribune*, but Nellie and her husband weren't keen on the idea. Kate never encouraged Bert to have a career in the newspaper business. Rather, she would fund his travels in Europe to keep him away from getting in the way of Medill's succession to the newspaper throne. Nellie and Robert wouldn't have minded their son Joe taking an interest in the *Chicago Tribune*, but he was more interested in the theatre.

Bert was active in the Young Republican Club at Yale, and in his final year he was elected president of the Scroll and Key Society with the help of Medill and Joe. Home from Yale in the summer of 1903, Bert lived in a McCormick mansion on Erie Street. When in the city, Kate lived up the street in the Joseph Medill mansion on Cass Street, later renamed Wabash Avenue, but she put it up for sale in 1903, perhaps to keep Bert out of his grandfather's house. Bert enrolled in Northwestern University law school and the family attorney, William Beale, arranged for him to volunteer at Isham, Lincoln and Beale, the city's premier law firm of the day. In early 1904, Bert traveled to Panama and observed the remnants of the French attempt at canal building; he idolized Teddy Roosevelt for taking on the building of the canal by Americans.

Republican boss Fred Busse approached Robert Patterson and Bert with an offer to nominate Bert for alderman to run against James Aloysius Quinn in the North Side silk stocking ward. The *Tribune* was a supporter of Republican causes, Bert agreed, campaigned hard, and was elected to the city council in April 1904 representing the 21st Ward. He also lacked appreciation for the Republican old guard—Senator Lorimer, Speaker Cannon, even Uncle Robert Patterson. Bert was a popular alderman, supporting popular causes and opposing corruption and entrenched politicians. He was his own man.

April 1905 also saw Edward F. Dunne become mayor of the City, promising to convert private traction companies to public ownership. Bert, on the Local Transportation Committee, opposed public ownership but favored stronger regulation. Although criticized by the press, his legislation promising lower fares and better service survived a mayoral veto. During a committee meeting, Bert was called out by Fred Busse and informed that he was being slated for the District president in the fall 1905 election. Bert felt the draw of higher elective office because he sensed that in the increasingly Democratic 21st Ward, he wasn't assured for reelection. Public and press criticism of current District leadership and the popularity of Bert helped him win, even though he declined to debate Frank Wenter, past District board president and the current Democratic candidate for president.

In 1905, while Bert was a bachelor in his mid-twenties, he began a relationship with a married woman, Amy, who was eight years his senior and the wife of Ed Adams, Bert's cousin. Ed's passion for alcohol exceeded his success in the brokerage business and Bert, appearing as a close friend, was discreet in his courtship of Amy. Bert helped Amy with the family budget when Ed's earnings fell short. The tall and muscular Bert towered over the diminutive and slightly stout Amy. Who would have expected anything was out of order?

Even under reform Mayor Dunne, corruption and crime were the highlights of government and politics in the city. Upton Sinclair's 1906 novel, *The Jungle*, exposed the exploits of the meat packing industry and gross pollution of the South Fork, aka Bubbly Creek. Its gaseous emissions were from the fermenting animal carcass waste dumped in the creek. The District had no authority over the packers, but Bert, now that he was president, used his office to engage the District engineers to devise means to sweeten the creek. Flushing the Stock Yard Slip with the combination of sewage and lake water, pumped from the Thirty-Ninth Street Pumping Station through the Thirty-Ninth Street conduit and into the slip at the Halsted Street outfall, made the slip and South Fork less of a nuisance. Bert was able to influence business leaders to place screens on the outfalls from the packing houses and to contribute toward dredging the slip.

In early 1907, Medill was brought back from Cleveland to begin his takeover of the *Tribune*, and Bert focused his attention on the District's Lockport powerhouse and developing the District as an electric utility, offering low-cost electricity in competition to Samuel Insull at Commonwealth Edison Company (ComEd). Mayor Busse and Bert set about to remake the lakefront by improving the Calumet and Chicago Harbors. Bert saw the Calumet Harbor as a way to implement the long-stalled Calumet-Sag Channel and get the federal government to pay for it. Riparian land owners opposed and stopped the project; the District was not to be in the harbor business. Bert shot back at anyone criticizing District projects, as he needed enemies to serve as targets for his aggressive energies.

Bert was enthusiastic about water power and the electrical department at the District. However, he was unable to convince the City to buy electricity from the District. He was outraged that the City put limits on the District for setting poles and stringing wires along the city streets, requirements not put on ComEd. Other newspapers were critical of ComEd, but not the *Tribune* because William Beale was friends with Samuel Insull. Mr. Insull was a pioneer in developing the generation and distribution of electrical energy and was able to lower the cost of electricity and out-compete the District for customers. Eventually, the City executed an agreement with the District to accept electricity for street lights, city hall, pumping stations, etc.

Bert found pleasure in challenging the dominance of ComEd as a supplier of electricity. He found that the Illinois & Michigan Canal commissioners had leased its plum water power development to the Economy Light and Power Company of Joliet (Economy), a company that was eventually acquired by Insull. Bert used his combative and persuasive powers to help defeat a congressman and three state legislators in the Joliet area who were favorites of ComEd. He tried to use the power of the District in other ways to thwart ComEd, like using eminent domain to acquire Economy assets in the Joliet region. He championed the cause of the deep waterway to block ComEd from developing water power on the Illinois River. Voters supported a state bond referendum in November 1908 to fund the construction of dams and locks from Lockport to Utica, a project eventually successful but delayed by financial conditions. Bert wanted the District to build a hydroelectric power plant south of Joliet, but was blocked by legal action.

Early in 1909, seeking opportunities beyond the District, Bert became the unpaid treasurer of the *Tribune* at the suggestion of Robert Patterson, an unexpected invitation, and he maintained an office on the 13th floor of the Tribune building. He used this opportunity to learn all he could about the business of running a newspaper. Robert Patterson and Kate were not on good terms, and Robert delighted in pushing Kate's less-favored son at the *Tribune* to aggravate Kate's plotting to have her favored son, Medill, as the heir to the *Tribune*. In March 1909, Joe Patterson's life in the theatre was not successful and he was without funds. He was brought into the *Tribune* by his father as a means of survival.

A buyout of the *Chicago Tribune* by Hearst, a rival publisher, was threatened. Kate and Nellie were surprisingly agreeable, but Bert and Joe fought back. Kate encouraged Medill to exert his authority, but the *Tribune* board of directors would not meet Medill's demands for authority and salary, causing a breakdown. Medill left the *Tribune*. Robert left for Philadelphia to attend to his dying mother; she died and he committed suicide. Bert and Joe gained control, convinced Kate and Nellie to keep the paper in the family, and began to gather support among the board of directors. Meanwhile, in the personal side of Bert's life, he finally became intimate with Amy in August 1909. He had been helping the Adams couple with funds to support their town apartment and country farm. The affair with Amy eventually became public, subjecting Bert to the wrath of Kate and causing him to become a social outcast.

Despite being nominated for another term as District trustee, not president as the law had changed, Bert traveled to Europe with Kate to seal a *Tribune* deal. While there, he visited a canal in England and a sewage treatment plant in Germany so the District could pay his expenses. Meanwhile, the *Tribune* exposed payoffs for votes to elect Republican Senator Lorimer and other party scandals leading to a rift between the *Tribune* and Republicans. The *Tribune* exposure didn't help Bert's campaign for reelection, which was faltering anyway. The campaign was harsh, focusing on Bert's interest in the *Tribune* and competition with ComEd. He was criticized for not focusing on public health and pushing the Calumet-Sag Channel project. Despite vigorous campaigning, he lost to Thomas A. Smyth, a former and respected president of the District board. Bert's final deed at the District was ribbon cutting for the North Shore Channel, even though it wasn't completed.

In October 1910, a newspaper price war broke out, an effort again by Hearst to break the *Tribune*, but the attempt failed. The *Tribun*e not only survived, but surged ahead in circulation due to shrewd maneuvering by the managing editor. In March 1911, Bert was named by the board of directors as acting president, pending the recovery of Medill. He moved to the fourth-floor office once occupied by Robert Patterson and dived into improving employee benefits, which became the envy of other newspapers' employees. He also moved the *Tribune* into the newsprint business, purchasing forest land and a paper mill in Canada, giving the newspaper dependable access to quality newsprint at controllable costs.

Bert's support of Teddy Roosevelt over William Howard Taft in the Republican Party in 1912 proved unsuccessful, yet he was instrumental in winning a primary election for a presidential candidate. Teddy Roosevelt won the primary in the city dominated by the Tribune, but was overwhelmed by the downstate vote for Taft. Circulation-boosting efforts elevated the newspaper above the competition, leaving even the Hearst *Examiner* far behind. Bert devoted himself to the newspaper business, gathering more admiration from the board of directors.

In 1914, the divorce of Amy and Ed was final, and Kate threatened to disown Bert, but she couldn't oust him from the *Tribune*; the directors thought too highly of his success and the profitability of the newspaper.

Kate and Nellie eventually relinquished their interest in the newspaper, and Bert signed a pact with Joe to manage the affairs of the *Tribune* until the two cousins were dead. Joe died in 1946, six years before Bert's death. Amy and Bert eventually were wed; Bert outlived her and went on to have a second wife. At Bert's insistence, the *Tribune* published its infamous headline *DEWEY BEATS TRUMAN* in November 1948. Bert remained at the helm of the *Tribune* until 1955, long enough to see another Republican, Eisenhower, in the White House.

References

Brown, George P. *Drainage Channel and Waterway: A History of the Effort to Secure an Effective and Harmless Method for the Disposal of the Sewage of the City of Chicago, and to Create a Navigable Channel Between Lake Michigan and the Mississippi River*. Chicago: R.R. Donnelley & Sons Company, 1894.

Chicago Tribune Archives. archives.chicagotribune.com.

Circuit Court of Cook County, Illinois, Chancery Division. James Abec, et al., Plaintiffs v. Metropolitan Water Reclamation District of Greater Chicago, Defendant. Case No. 03 CH 21800 consolidated with Case No. 04 CH 752 and 03 MI 718897.

The Colonel Robert R. McCormick Research Center at Cantigny Park. Robert R. McCormick Sanitary District of Chicago files. Wheaton, IL.

Dowell, Amy, et al. *150 Years of Engineering Excellence*. Palatine, IL: American Society of Civil Engineering Illinois Section, 2003.

Hogan, Herman. *The First Century: The Chicago Bar Association 1874–1974*. Chicago: Rand McNally & Company, 1974.

Illinois General Assembly. "Previous General Assemblies." ilga.gov/previousga.asp

Lux, Lawrence E., et al. *A History of Public Works in Metropolitan Chicago*. Kansas City, MO: American Public Works Association, 2008.

Metropolitan Sanitary District of Greater Chicago Proceedings of the Board of Trustees, 1955 and 1975.

Metropolitan Water Reclamation District of Greater Chicago. Proceedings of the Board of Commissioners, 1988 and 1989.

New York Times Archives. nytimes.com/ref/membercemtermutarchive.html.

Piersen, Joe. *C&NW Lines North of Mayfair: Maps*. Deerfield, IL: Chicago & North Western Historical Society, 2004.

ravenswoodmanor.com/centennial/history.

Rogers Park/West Ridge Historical Society. *The Historian*. (Summer 2014): 12–13.

Smith, Richard Norton. *The Colonel: The Life and Legend of Robert R. McCormick, 1880–1955*. Boston: Houghton Mifflin Company, 1997.

Supreme Court of Illinois. *In re Information to Discipline Certain Attorneys of the Sanitary District of Chicago*, No. 20357, 351 Ill. 206, 184 N.E. 332, February 23, 1933.

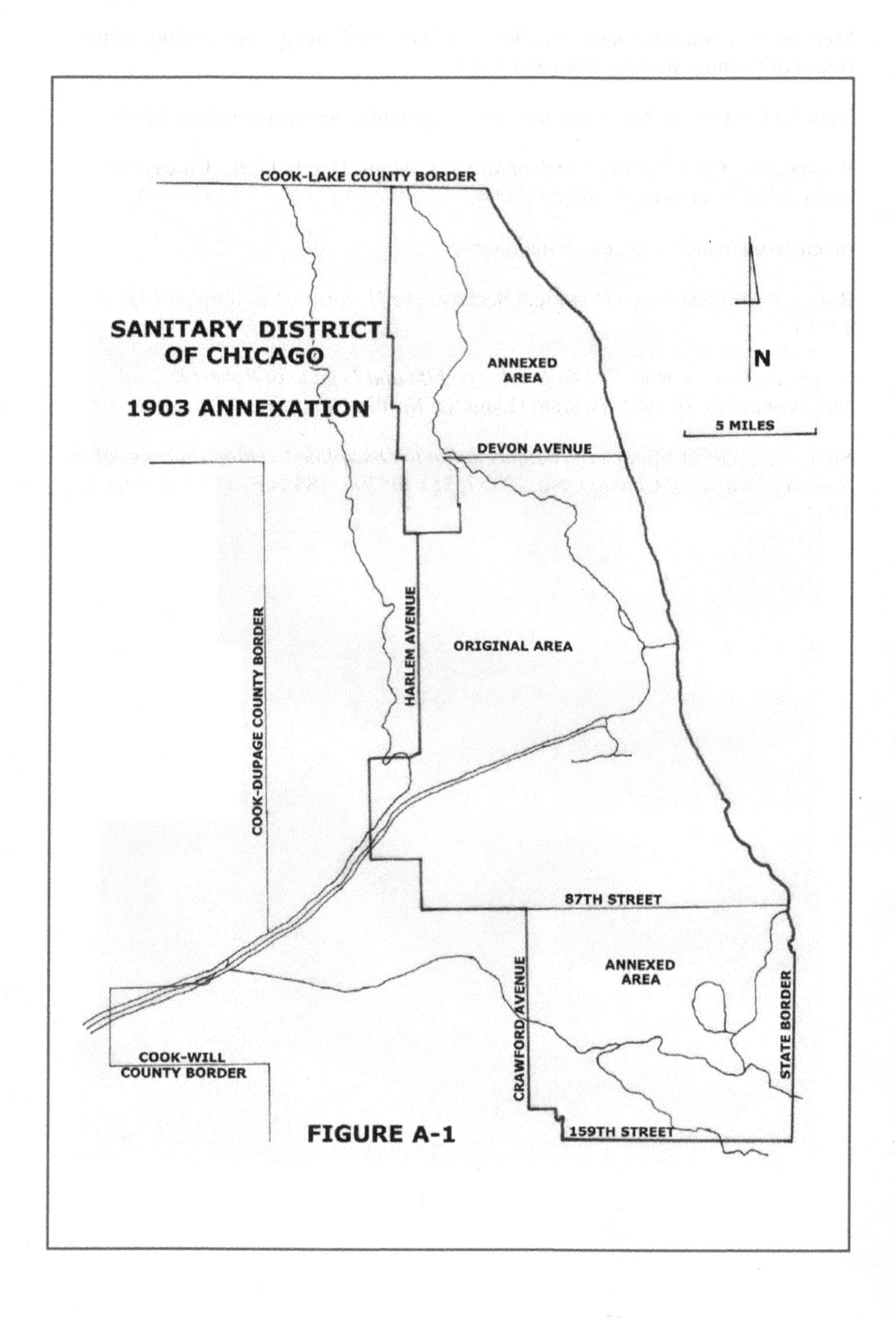
COOK-LAKE COUNTY BORDER
SANITARY DISTRICT
OF CHICAGO
1903 ANNEXATION
ANNEXED
AREA
N
5 MILES
DEVON AVENUE
HARLEM AVENUE
ORIGINAL AREA
COOK-DUPAGE COUNTY BORDER
87TH STREET
ANNEXED
AREA
CRAWFORD AVENUE
STATE BORDER
COOK-WILL
COUNTY BORDER
159TH STREET
FIGURE A-1

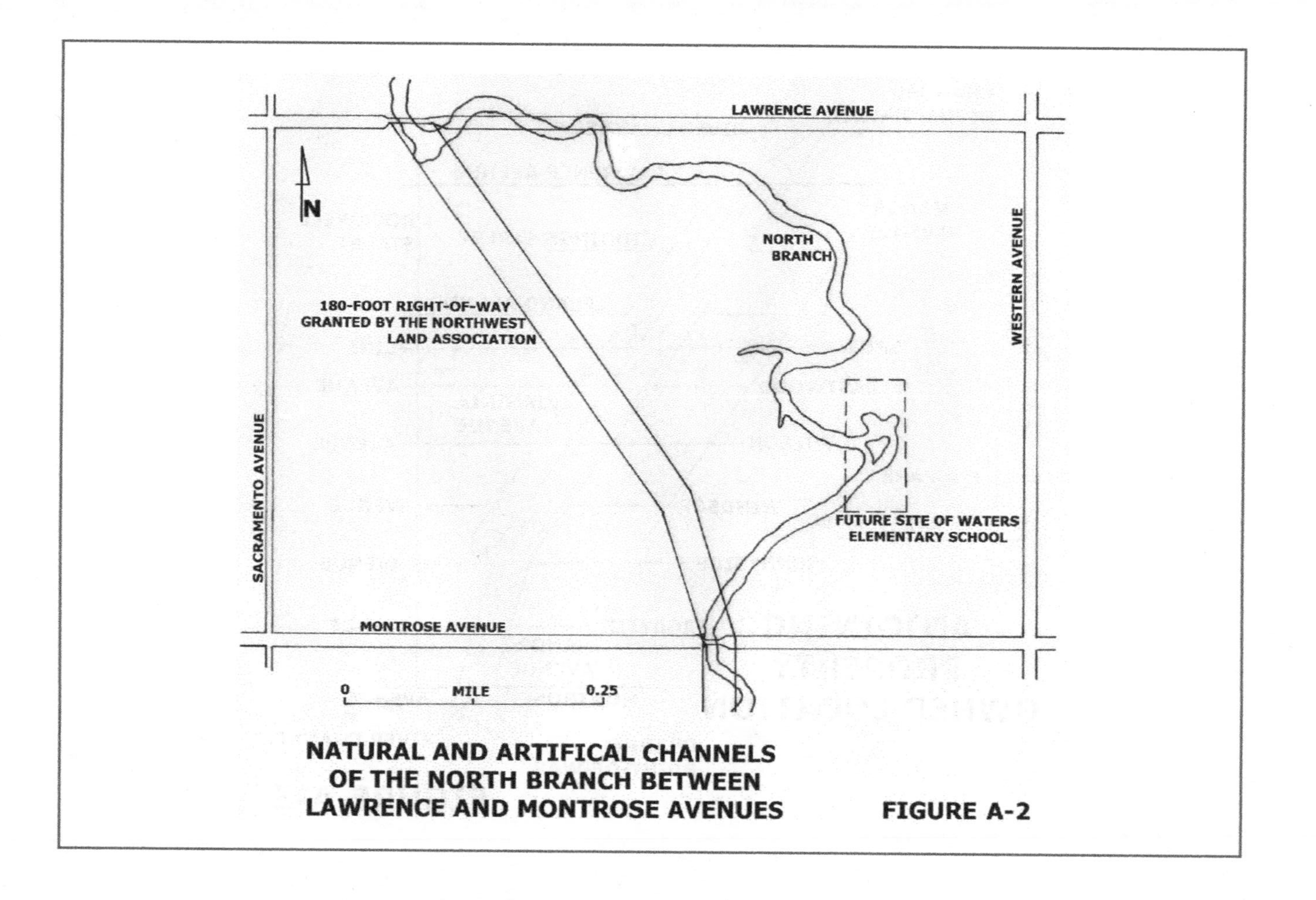

NATURAL AND ARTIFICAL CHANNELS OF THE NORTH BRANCH BETWEEN LAWRENCE AND MONTROSE AVENUES

FIGURE A-2

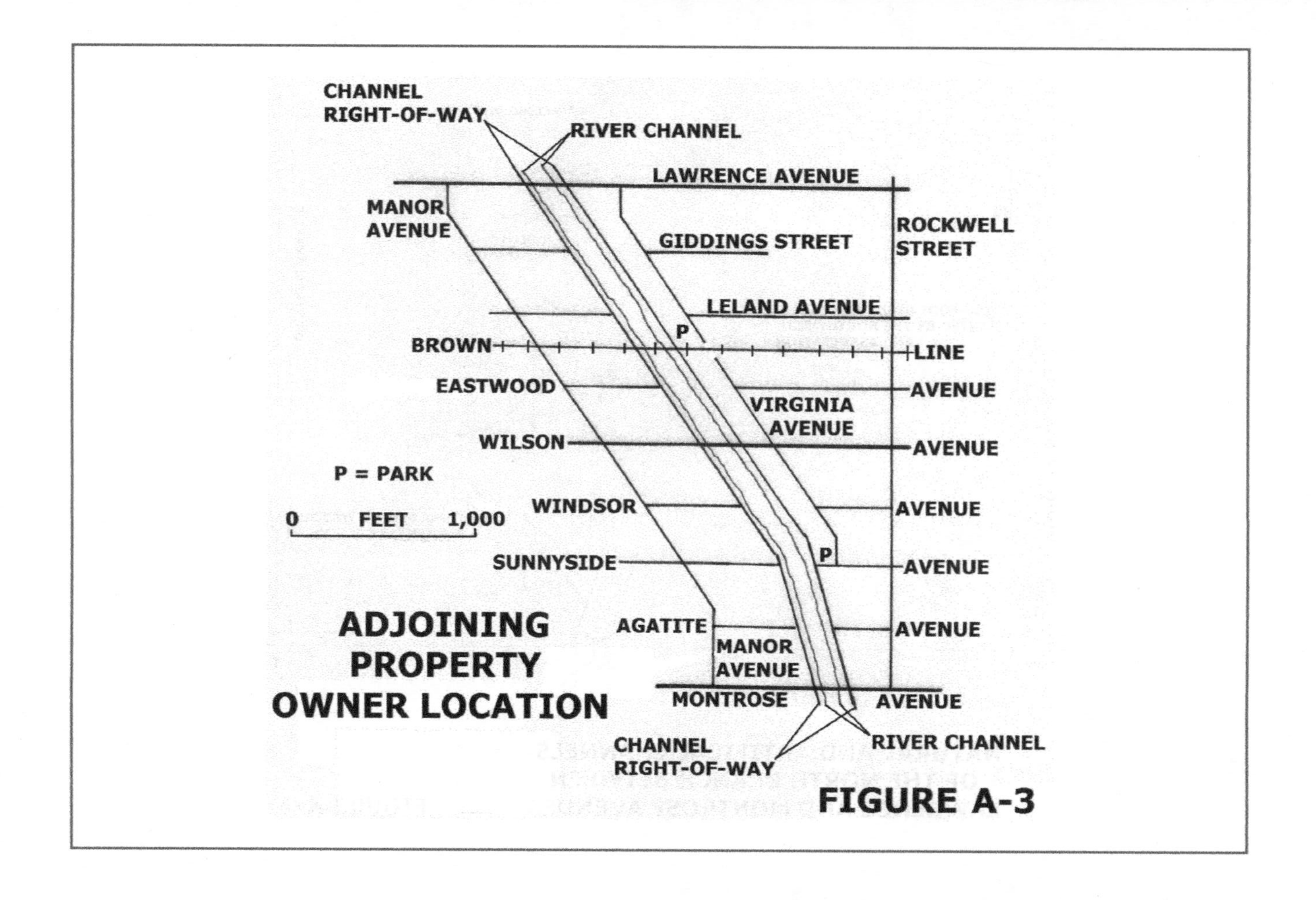
CHANNEL
RIGHT-OF-WAY
RIVER CHANNEL
LAWRENCE AVENUE
MANOR
AVENUE
GIDDINGS STREET
ROCKWELL
STREET
LELAND AVENUE
P
BROWN
LINE
EASTWOOD
AVENUE
VIRGINIA
AVENUE
WILSON
AVENUE
P = PARK
WINDSOR
AVENUE
0 FEET 1,000
SUNNYSIDE
P
AVENUE
ADJOINING
PROPERTY
OWNER LOCATION
AGATITE
AVENUE
MANOR
AVENUE
MONTROSE
AVENUE
CHANNEL
RIGHT-OF-WAY
RIVER CHANNEL

FIGURE A-3

Photograph A.1: The route of the proposed McCormick Road lies to the right of the row of transmission towers in this July 11, 1924, north-facing view from a point midway between Howard and Oakton Streets. Considerable spoil remains in the distance. At left is the rail siding entrance to the North Side Sewage Treatment Works construction, and the manufactured gas plant and holding tank is in the left background. The Mayfair branch railroad is at right with the Weber yard in the right background. (MWRD photo 10872)

Photograph A.2: Two laborers are using a wooden plank to set the grade of the sub-base material while a third is using a tamper to compact the sub-base material for the west ten-foot slab on July 9, 1925. The four-lane pavement design was thought to be advanced for its time: two ten-foot wide reinforced Portland cement concrete slabs for the outside lanes intended for truck traffic, flanking two center ten-foot-wide lanes intended for automobiles paved with asphalt concrete. Curbs, gutters, storm inlets, storm sewers draining to the channel, and electric lighting completed the improvement. (MWRD photo 11898)

Photograph A.3: Looking east at what will become the intersection with McCormick Road on July 15, 1925. A flagman is controlling traffic on Dempster Street while paving of McCormick Road proceeds. Most of the local cross streets in unincorporated areas were unpaved at the time. Dempster was one of the more travelled and better maintained. (MWRD photo 11936)

Photograph A.4: July 15, 1925. A hopper at right is being lifted to dump its load of materials into the Koehring concrete mixer, while the labors at left are pouring the previous batch of concrete into the prepared pavement form with reinforcing bars already in place. The hoppers were loaded at the contractors batching plant near Oakton Street with the correct amount of aggregate, cement, and sand for one mixer batch of concrete. Several hoppers were towed by a locomotive to the paving site. (MWRD photo 11940)

Photograph A.5: August 17, 1925. Looking southwest from the railroad embankment, the northern end of McCormick Road intersects with West Railroad Avenue, presently Green Bay Road, in Evanston. The paving of both truck lanes with use of the Koehring concrete mixer has been completed. The intersection will also be paved with reinforced concrete before paving the two center lanes with asphalt begins. (MWRD photo 12073)

Photograph A.6: January 16, 1926. Excavation for the northern approach to the Chicago & North Western Railroad Mayfair Division viaduct over McCormick Road is delayed by flooding of the cut. Work on the diversion trestles and track on the far side of the railroad is underway. The bridges for the elevated electric railroad crossing of the channel, Mayfair Division, and McCormick Road were recently completed. This southeast-facing view was taken from the gas holding tank on Oakton Street. The large square structure in the background faces Howard Street and is a greenhouse built by Budlong Farms in 1925 and still in operation in 1945. The property was subsequently sold to Hibbard, Spencer & Bartlett and a large warehouse replaced the greenhouse. (MWRD photo 12215)

Photograph A.7: A long freight train is passing on the diverted track on July 7, 1926, while pillars and spans are being formed and cast with reinforced concrete for the piers of the railroad viaduct over McCormick Road. Excavation of the northern approach proceeds beyond the freight train. The North Shore Channel is at right and the manufactured gas plant on Oakton Street is at left. (MWRD photo 12790)

Photograph A.8: Structural steel girders are being set in place on top of the three parallel viaduct piers on August 10, 1926, while rough grading on the northern approach proceeds in the foreground. The abutments of the viaduct flanked the two outside piers, allowing for a sidewalk on each side of the roadway. (MWRD photo 12897)

Photograph A.9: October 8, 1927. Southbound vehicles passing through the viaduct on McCormick Road, which has been open to traffic for over a year. The door on the small building on the east abutment enters a small pumping station for keeping the viaduct dry during rain events when the channel rises too high for gravity drainage. (MWRD photo 13491)

Photograph A.10: McCormick Road had become a popular thoroughfare for the city's North Side and the northern suburbs of Evanston and Skokie by the time this photo was taken on July 6, 1928. With the handsome lighting, it soon became known as McCormick Boulevard. The granite monument for the roadway is at right and beyond it in the background is an excavating machine removing excess spoil from District property for the creation of a bridle path. (MWRD photo 14119)

Photograph A.11: A great deal of effort went into removing the remaining spoil along the North Shore Channel to make the vista more appealing to drivers on McCormick Road and to respond to requests by nearby municipalities for recreational space. Perhaps the photographer was unaware on July 17, 1928, that this subject would result in trouble for officials at the District and their contractors. (MWRD photo 14176)

Photograph A.12: A section of land south of Howard Street already cleared of excess spoil, and a strip dressed with a coating of cinders, are already being used by equestrians on July 17, 1928. Even the cinders contributed to the trouble at the District and the popularized the name of the payoffs to contractors: *The Million Dollar Bridle Path.* (MWRD photo 14177)

Photograph A.13: A bridle path was not the only reason for spoil removal. The National Brick Company continued its obligation to remove excess clay spoil for the manufacture of bricks at its plant south of Howard Street. Shown here, a tug and barge pass Howard Street on July 17, 1928, on its way to the company's nearby dock. The familiar gas holding tank on Oakton Street is in center background, and the District's North Side Sewage Treatment Works, presently the O'Brien Water Reclamation Plant, Pump and Blower Building, is at left. The outfall on the west bank is part of Skokie's Howard Street sewer. (MWRD photo 14172)

Photograph A.14: Ellis Sylvester Chesbrough, chief engineer of the Board of Sewerage Commissioners in 1855 and later the Commissioner of Public Works, oversaw development of sewer and water infrastructure for Chicago for 25 years. He was the first engineer to suggest a large canal to reverse the flow of the Chicago River to protect the water supply. (Photo from the internet)

Photograph A.15: Isham Randolph, chief engineer of the Sanitary District of Chicago from 1893 to 1907, successfully oversaw and administered construction of the original Sanitary & Ship Canal, the canal extension with lock and powerhouse, North Branch improvement, South Branch widening, and several bridge replacements. (MWRD photo 10384)

Photograph A.16: Robert R. McCormick, president of the Sanitary District of Chicago from 1905 to 1910, was elected at age 25 in a landslide that brought a Republican majority to the board of trustees. Bert, as he was known before earning *The Colonel* moniker in World War I, claimed to be a reformer and tried to run the District as if it was his personal business. Under his reign, the Calumet-Sag Channel was sidelined in favor of making the District an electric utility with cheap hydroelectric power in competition with Samuel Insull at Commonwealth Edison. He didn't succeed and lost his bid for reelection in 1910. (Photo from a newspaper advertisement on April 28, 1907, in the files of The Robert R. McCormick Research Center at Cantigny Park, Wheaton, Illinois)

Acknowledgments

I'm grateful for assistance and encouragement received from many in my research for this book. Many—too many to mention—former and current members of the staff in the departments of General Administration, Engineering, Law, and Maintenance & Operations at the Metropolitan Water Reclamation District have assisted in finding, supplying, and clarifying factual details. It makes me proud to have been and continue to be associated with this organization. The District has racked up over 125 years of service, a service vital every day to the health and wellbeing of the people served and the environment in which we live.

Additional assistance was received from staff members or volunteers at the following: Chicago Public Library Municipal Reference Collection, Chicago Public Library Sulzer Branch, Colonel Robert R. McCormick Research Center at Cantigny Park, Evanston Historical Museum, Evanston Utilities Department, Illinois Department of Transportation District 1, Rogers Park/West Ridge Historical Society, Sargent & Lundy, Wilmette Harbor Association, and Wilmette Public Library.

My family is and has been continuously supportive and patient with my preoccupation in research and writing. For years, Marsha endured my occupying an alcove in our bedroom as an office and then oversaw the conversion of an unused bedroom for a proper office, for which I'm very grateful. I'm so fortunate to have Marsha as companion and friend and ever thankful for that fateful day when her Maverick wouldn't start.

I'm indebted to Sharon Woodhouse at Conspire Creative and the excellent team she assembled to design, edit, and produce this book. Her intimacy and interest in Chicago, enthusiasm for producing good books, expertise in all phases of publishing and promoting, and her direct and common sense approach to business is deeply appreciated. Working with her is easy, enjoyable, and rewarding.

During the preparation of this manuscript I was involved in three other projects that enriched my experience and deepened my appreciation. Richard Cahan and Michael Williams graciously listed me as first acknowledgment in *The Lost Panoramas*. Working with them increased my awareness of the value of the District's photographic archives and added to my understanding of the book publishing experience. Artist Lindsay Olson in Oak Park took on the artistic interpretation of wastewater treatment in a series of works called *Manufactured River*. I was pleased to be her technical tutor and gain appreciation of how the engineering in wastewater looks from an artist's point of view. Dancer and teacher Clare Tallon Ruen invited me to work with her in *Pipes & Precipitation*, a grant funded project to enrich the experience of sixth and third grade students in the Evanston/Skokie School District 65. I found that I was better at communicating with adults than children and young adults, realized more how much work goes into teaching and how much teachers are valued by students eager to learn, and learned how interpretive dance can help students understand water and wastewater treatment processes.

Index

Page numbers in italics refer to photographs and captions. The letter "t" following a page number denotes a table; the letter "f" following a page number denotes a figure.

C

D

E

F

G

H

I

M

O

P

Q

R

T

U

V

W

Y

About the Author

Richard "Dick" Lanyon has had a life-long association with the waterways in and around Chicago. He grew up along the North Branch, attended the University of Illinois Navy Pier campus, worked as a beginning engineer on the Lake Diversion legal controversy, and capped his working life with a 48-year run with the Metropolitan Water Reclamation District. Lanyon retired as executive director of the MWRD in 2010, a position he held for 4.5 years.

As Executive Director, Dick directed the day-to-day operations of the MWRD, which included 2,100 employees serving five million people in Cook County and the industrial equivalent of another four million people.

His first book, *Building the Canal to Save Chicago* (2012), received the 2013 Abel Wolman Award from the Chicago Metro Chapter of the American Public Works Association Chicago Metro Chapter for best new book in public works history.

Other awards include: American Society of Civil Engineer's National Government Civil Engineer of the Year Award in 1999, Distinguished Alumnus of the Department of Civil and Environmental Engineering at the University of Illinois at Urbana-Champaign in 2003, Edward J. Cleary Award from the American Academy of Environmental Engineers and Scientists in 2011, and Distinguished Service Award from the National Association of Clean Water Agencies (NACWA) in 2011. Lanyon is a past president of the Illinois Section of the American Society of Civil Engineers and holds Bachelors and Masters of Civil Engineering degrees from UIUC. In 2013, he was inducted into the NACWA (National Association of Clean Water Agencies) Hall of Fame.

Lanyon has been involved in a variety of technical activities for the above and other organizations, and he has served in a number of leadership roles on environmental protection and water resource management matters for federal, state, and local agencies and organizations. Dick served on the Evanston Public Library Board of Directors and as alderman of the 8th Ward on the Evanston City Council. He is currently the Chairman of the Evanston Utilities Commission.

He enjoys biking along the waterways and Lake Michigan, and through the neighborhoods of his hometown Evanston, Illinois, where he lives with his wife, Marsha Richman.

THE MWRD PHOTO ARCHIVE

One of Dick's voluntary retirement projects is creating a catalog of the Metropolitan Water Reclamation District's archive of 14,000 glass-plate negatives documenting District construction activities over the period 1894 through 1930. The archive brings alive the early work of the District and landscapes of the growing metropolis. In his first book *Building the Canal to Save Chicago*. Dick used 180 of these photos and 183 are used in the current book. In *The Lost Panoramas*, by Richard Cahan and Michael Williams, 200 of these photographs were used by the authors with Dick's assistance in identification and explanation. Dick provides advice and guidance to anyone with questions about water in the Chicago area and the history of the District and its photographic archive.